Living with Nature in the Anthropobscene

Rod Giblett

Living with Nature in the Anthropobscene

Literature, Philosophy, and Political Ecology

palgrave
macmillan

Rod Giblett
Deakin University
Coburg, VIC, Australia

ISBN 978-3-032-12821-8 ISBN 978-3-032-12822-5 (eBook)
https://doi.org/10.1007/978-3-032-12822-5

This Palgrave Macmillan imprint is published by the registered company Springer Nature Switzerland AG.
The registered company address is: Gewerbestrasse 11, 6330 Cham, Switzerland

Dedicated to keeping alive the memory of Raymond Williams (1921–1988),
the founder of ecocultural studies and "a new kind of political ecology"
for living with nature in the Anthropobscene

Competing Interests The author has no competing interests to declare that are relevant to the content of this manuscript.

Contents

CHAPTER 1

Introduction to Philosophy, Ecoculture, and Political Ecology for Living with Nature in the Anthropobscene

Abstract The Anthropobscene is defined as the laying bare of industrial products and wastes in the land, air, water, and the bodies of living beings; the heating up of land, air, and water in the new, disrupted (and disruptive) arrangement of the four elements and the four seasons in nature everywhere; and the impacts on them that were supposed to remain hidden, out of sight, and out of mind. This laying bare entails environmental, biological, mental, spiritual, cultural, and social effects and consequences in and for all living beings and their/our earthly home of "the vicious exploitative actions of corporations, governments and other agencies" in the hyper-industrial, computational, and psychogeopathological capitalism of today. Nature is defined as the processes and places of land, air, and water, their plants and animals (including humans), the provision of life-giving air and energy, the temporal rhythms of the day and the seasons, and the aerial vagaries of the weather and climate. Industrial wastes have changed climates, disrupted seasonal cycles, and exacerbated weather extremes producing mega-fires and -floods, destructive droughts and storms, and the hottest day, month or year on record. This introductory chapter also defines philosophy and ecoculture, discusses political ecology, and outlines the chapters that follow.

In a time of climate collapse, the ongoing extinction of many species, and the end of the age of fossil fuels, this book proposes ways of living with nature locally, globally, and psycho-symbiotically. The present era of the

R. Giblett, *Living with Nature in the Anthropobscene*,
https://doi.org/10.1007/978-3-032-12822-5_1

Anthropobscene is the laying bare of industrial products and wastes in the land, air, water, and the bodies of living beings; the heating up of land, air, and water in the new, disrupted (and disruptive) arrangement of the four elements and the supposedly four seasons; and the impacts on them that were meant to remain hidden, out of sight and out of mind. Prime examples of the Anthropobscene are the slimy mud hell of World War I (WWI) and the sublime bombed cities of World War II (WWII; as we will see in Chaps. 7 and 8). The laying bare of what is supposed to remain out of sight and out of mind is the obscene of the Anthropobscene. It is not an epoch, a 'cene' like the Anthropocene with a contested beginning, but a psychopolitical and ecological process that has its roots in the ancient Greek philosophy of the four elements and humors and in the ancient Roman philosophy of the sublime (as we will see in Chaps. 2 and 3). The Anthropocene is the epoch of the laying down in the geological strata of industrially anthropogenic products and wastes in the past and the present and in the archaeological record for the future (Crutzen 2006).

In 2018 I thought I was the first to come up with the term 'Anthropobscene.' I introduced it to the world publicly in a lecture in April 2019 at Susquehanna University in Selinsgrove, PA (Giblett 2018). A reviewer of the manuscript of this book pointed out to me that, unbeknown to me at the time, Jussi Parikka had already coined the term in 2013. Parikka stated:

> this portmanteau word [Anthropobscene] combines Anthropocene with obscene, thus highlighting the vicious exploitative actions of corporations, governments and other agencies operating on different levels: from human individuals to multigovernmental organisations and transnational corporations. (Bishop and Parikka 2013)

I concur with this quasi definition and add extra dimensions to it: the Anthropobscene is the laying bare in the air, land, water, and land of the environmental, biological, mental, spiritual, cultural and social effects and consequences in and for all living beings and their/our earthly home of "the vicious exploitative actions of corporations, governments and other agencies" in the hyper-industrial, computational, and psychogeopathological capitalism of today.

A Google search for Anthropobscene in 2018 and again in 2025 did not yield the result of Parikka's coinage. Google always searched instead for 'Anthropocene.' Along the same lines, one of the senior academics at Susquehanna University who proofread the proposed title and abstract for my lecture thought Anthropobscene was a typographical error, perhaps like Google. I doubt if they attended my lecture. Similarly an editor for a prominent series of books of ecocriticism published by an eminent academic publisher thought that Anthropobscene was "surely a typo" and subsequently rejected the book proposal for *Living with Nature in the Anthopobscene*. I doubt if the coiners of Anthropocene suffered the same indignities. Critiques initially of Anthropocene were directed at the validity of the concept and its pertinence to describe the current geological epoch, not to the coinage *per se* and the act of making it. This difference points to a fundamental divide between the sciences and the humanities, and the reception of new work in them, between the sciences' discursive construction of concepts and objects, and the humanities' critical analysis of discourses, cultural processes, and subject positions.[1]

Rather than merely laying down strata in the geological record of the Anthropocene, the Anthropobscene is laying bare industrial products and wastes in the present, in nature everywhere, in the land, air, water, and the minds and bodies of living beings. The Anthropobscene came out of the economic politics of mercantile capitalism, industrial capitalism, enclosure of the commons, private property, the commodity market, and the feral quaking zone of industrial wastelands (depicted on the cover of *Living with Nature in the Anthropobscene*). I define nature as the processes and places of land, air, and water, their plants and animals (including humans), the provision of life-giving air and energy, the temporal rhythms of the day and the seasons, and the aerial vagaries of the weather and climate. Industrial wastes have changed climates, disrupted seasonal cycles, and exacerbated weather extremes producing mega-fires and -floods, destructive droughts and storms, and the hottest day, month or year on record. All of these impact on human and more-than-human life on earth with devastating and deadly consequences for all lives.

Living with Nature in the Anthropobscene brings together philosophy, ecoculture, and political ecology for living with nature locally, globally, and psycho-symbiotically. I define philosophy as thinking creatively and critically through concepts, ideas, symbols, and figures, such as metaphors. I define ecoculture as the inter-relations between human beings,

[1] For further discussion of this difference in relation to natural history and nature writing drawing on the work of Michel Foucault, see Giblett (2026a, chapter 1). For further discussion of the transdisciplinary environmental humanities crossing the great divide between 'the two cultures' of the sciences and the humanities, see Giblett (2011, chapter 1).

non-human beings, and the earthly household (*oikis*) enacted and expressed in cultivation (working the land, dry, and wet, for physical sustenance, such as agriculture—cultivation of drylands—and paludiculture—cultivation of wetlands), artefacts, arts, the body, the city, storytelling, writing, mythology, religion, etc., (nurturing the land for physical and spiritual sustenance, for environ-mental health).

Political ecology is a catch-all category that includes femo- and psycho-political ecology, as well as Raymond Williams's "a new kind of political ecology" and "socialist ecology," Felix Guattari's "generalized ecology" and Paul Virilio's "grey ecology" (Giblett 2025, 128–130). Psychopolitical ecology critiques the politics of the psyche (mind, spirit), the psychodynamics, including psychopathologies, of politics, and both their inter-relations with the earthly household of nature from psychogeopathology to psychogeosymbiosis. These inter-relations include investments of desire and capital, yields of pleasure and profits, and relations of power or love. Psychopolitical ecology comes from bringing together in fruitful dialogue and historical synthesis the work of Walter Benjamin (1892–1940) and Henry David Thoreau (1817–1862) for waking to nature with them (Giblett 2025). Femopolitical ecology deconstructs and decolonises the gendered construction of reality, especially in patriarchy, not only of living beings, but also of processes, places, and spaces, and nurtures love for them. It comes from bringing together the words and artworks of many wise women for loving nature (Giblett 2026b).

Living with Nature in the Anthropobscene traces some of the intersections and interchanges and some aspects of the trilogue between western philosophy, ecoculture, and femo- and psycho-political ecology. Drawing on a rich variety of ecocultural texts and from a wide range of western philosophers and other thinkers from the ancient Greeks to the present day (principally Peter Sloterdijk and occasionally Byung-Chul Han), the book invites the reader to step along the pathway for living with nature sacrally in the Anthropobscene, moving away from the drive for mastery of nature to the desire for, and pleasures of, mutuality with the living earth.

Living with Nature in the Anthropobscene is a long, intermittent, respectful, and occasionally critical dialogue with Sloterdijk, the founder of psychopolitical spherology. It by no means addresses and encompasses all the myriad facets of his huge, diverse, suggestive, and immensely productive work, but engages with areas of consonance as indicated in the table of contents and critiques topics of dissonance, principally his adherence to the precepts of western philosophy and its project since its inception of the sublime and sublimation summed up neatly and critically by Karl Marx in their culmination in industrial capitalism as "all that is solid melts [or is sublimed] into air."[2] Psychopolitical ecology rebuts both philosophy and industrial capitalism and engages in desublimation in which all that is airy, or sublim(at)ed, "melts," or is desublim(at)ed via slime, into water, or liquid. The intimate relationship between the sublime and slime is encapsulated brilliantly by Zoë Sofoulis (1988) in her portmanteau "s(ub)lime" in which "slime is the secret of the sublime."[3]

Each chapter of *Living with Nature in the Anthropobscene* as outlined in detail below in Sect. 1.1 engages with an aspect of the work of Sloterdijk that provides the template for the discussion of the topics mentioned in the table of contents. An alternative set of abstracts for each chapter would have traced through *Living with Nature in the Anthropobscene* its engagement with Sloterdijk's work. A second book is contained, like a portmanteau, within the present one, thus presenting two books for the price of one and in the compass of one. *Living with Nature in the Anthropobscene* is the first book in English to engage across the board with philosophy, history, literature, ecology, theology, and religion in some of his work. It is by no means a comprehensive engagement with all these aspects of and in his work. *Living with Nature in the Anthropobscene* is also the second book of psychopolitical ecology and of femopolitical ecology (Giblett 2025, 2026b).

I am mindful and respectful that I stand on the shoulders of these and many other giants, including Raymond Williams, the founder of ecocultural studies and "a new kind of political ecology," hence the dedication of *Living with Nature in the Anthropobscene* to keeping his memory alive for living nature in the Anthropobscene whose destruction of the earth he was alert to, especially in his late essays and *Towards 2000* discussed in Chap. 11. If *Living with Nature in the Anthropobscene* reaches new heights, dives to new depths, deconstructs the hierarchy between them, and decolonises the power relations between them, it is because of these and other forerunners.

[2] "Melts" translates "*verdampft*" (cited in Sloterdijk 2025, 34 n1), usually translated as 'evaporated' or 'vaporised,' and so pertinent to liquids being transformed into air, or gas. Sublimed is more precise for solids being transformed into air, or gas.

[3] The sublime, sublimation, and slime are discussed further in Chaps. 2 and 3.

Living with Nature in the Anthropobscene is a transdisciplinary, historical, philosophical, political, ecocultural, ecocritical, theological, and religious study in the environmental humanities following in the footsteps of those at their forefront who have gone, or who are still going before. It invites the reader to step out of the Anthropocene and Anthropobscene in the past and present along the pathway towards living with nature sacrally and psycho-symbiotically in the present and future in what Glenn Albrecht (2019) calls the Symbiocene, the hope-for age superseding the Anthropobscene and Anthropocene. The Symbiocene is the paradigm of the desire for mutuality with the earth posed against the failed paradigm of the drive to mastery and the will to power of the anthropocentric over the earth (see Giblett 2011, especially Figure 2, 32–34). The Symbiocene can be defined as the geological epoch of the laying and layering up and down of bodies of land, waters, air, beings, and things in biogenic and non-biogenic strata, and the inter-corporeal relationship of loving union between them in the feral and native quaking zones. The Symbiocene comes out of the economic politics of the commons, compassion for all beings, mutual aid, the carnivalesque marketplace (not the capitalist market) and the native quaking zone of an indigenous wetland, as distinct from the feral quaking zone of an industrial wasteland. The Symbiocene comes out of its love for, and paradigm of, mutuality (see Giblett 2011, chapter 1).

1.1 The Way Ahead

The sublime in chemical terms transforms the solid, or earth, into the gaseous, or air. The four elements of earth, air, fire, and water are foundational to Greek philosophy and to western philosophy and culture more generally in the Anthropobscene. Chapter 2 of *Living with Nature in the Anthropobscene* brings western philosophy, ecoculture, and psychopolitical ecology together beginning with the classical Greek philosophy of the four elements. These elements for the Greek philosophers should stay in their proper place and not mix. If and when they do mix, they produce unhealthy water, bad air, and disease ('malaria'). Out of the elements, the Greek philosophy of monsters and the humors, or psychosomatic affects, also arose. Chapter 2 focusses on monsters from the ancient Greeks, through Grendel and his mother in the old English elegiac long poem *Beowulf*, to Satan in Milton's epic long poem *Paradise Lost*. It also focusses on the humor of melancholy and traces its history from the ancient Greeks through the seventeenth century with "the slough of despond" in John Bunyan's *Pilgrim's Progress*, and into nineteenth-century detective fiction and late twentieth-century cultural studies.

The Greek philosophy of the four elements is foundational to the classical Roman philosophy of the sublime and the sublimation of the body into the mind or spirit in the Anthropobscene. Chapter 3 of *Living with Nature in the Anthropobscene* brings western philosophy, ecoculture, and psycho- and femo-political ecology together beginning with the sublime and sublimation. The sublime for Zoë Sofoulis is always shadowed by slime, which she encapsulates in the parenthetical portmanteau s(ub)lime. Slime is the secret of the sublime. This chapter focusses on: the sublime in work of Longinus and slime in Dante's *Inferno*; the sublime in the work of Immanuel Kant and Edmund Burke; slime in Jean-Paul Sartre's *Being and Nothingness*; and the s(ub)lime in late twentieth-century critical theory, including Jean-François Lyotard's and Marshall Berman's accounts of modernity. The sublime is central to western aesthetics, especially in the modern period. Around the name of the sublime, Lyotard remarks, "modernity triumphed," not least over wetlands in the short term, by draining, filling, or embanking them, ironically using machines powered by fossil fuels from relic wetlands. The fossilised remains of swamps are sublim(at)ed to power machines and vehicles with steam and internal combustion engines in what Lewis Mumford (1934, 156–158) called "carboniferous capitalism" that gave rise to the age of global boiling and the Anthropobscene. This chapter critiques the gender politics and ecoculture of the masculinist sublime and the misogynist feminisation of slime.

The sublime and sublimation are always haunted and made possible by their other of slime and the uncanny. Chapter 4 of *Living with Nature in the Anthropobscene* brings western philosophy, ecoculture, and psychopolitical ecology together beginning with the German philosophy and psychology of the uncanny, the combination of fascination and horror. The uncanny for Sofoulis is associated with the obverse of the sublime. This chapter focusses on the uncanny in the work of Friederich Schelling that comes out of the philosophy of polytheism and mythology, the bases for monotheism and theology. At a time and for a world in which some sectors of the monotheistic religions are in aggressive, and sometimes violent, contention against those of other monotheistic religions, and even against other sectors within their own monotheistic religion, Schelling's discussion of the uncanny enables rethinking the relationship between mythology and monotheistic and polytheistic religions in a culturally and politically liberatory and progressive way in the Anthropobscene. This chapter makes a plea for polytheism in a world and age riven by conflicts between and within monotheistic religions. Polytheism is living with nature in the earthly household of the ecosphere. The uncanny is the means to do it. It taps into a timely and rich vein of thinking about the divine, darkness, light, aura, religion, mythology, and the monstrous that enables rethinking the inter-relationship between all of them in environmentally- and animal-friendly terms.

The uncanny was taken up by Sigmund Freud in his famous essay of 1919 on the topic. Chapter 5 of *Living with Nature in the Anthropobscene* brings western philosophy, ecoculture, and psycho- and femo-political ecology together beginning with the psychoanalytic theory of the uncanny. This chapter focusses on Freud's essay on the uncanny and its neglected eco-cultural aspects, especially his reading of what he calls "a naïve, but extraordinarily uncanny story" about crocodiles and a table with carved crocodile legs that he "came across" in *The Strand* magazine. This chapter argues that the uncanny is the return *to* the repressed that brings into the open in the present what is in the past and long forgotten and opens a pathway to hope in and the future. It also critiques the gender politics, ecoculture, and psycho- and femo- political ecology of a late twentieth-century Australian eco-feminist's recount of her encounter with a monstrous crocodile in an uncanny swamp. The chapter concludes by asking the question what is religion? It answers that religion is the uncanny without the boundaries of mere reason. The uncanny does not have reason and has absented itself from it. The uncanny swamp for Henry David Thoreau is religion in his definition as "where our love is." Religion is love for places and places that love, such as swamps for Thoreau in which he found hope in the Anthropobscene.

Besides the uncanny, Freud also took up melancholy and mourning in another famous essay on both topics. Chapter 6 of *Living with Nature in the Anthropobscene* brings western philosophy, ecoculture, and psychopolitical ecology together beginning with the psychoanalytic theory of melancholy and mourning. It focusses on Freud's essay on mourning and melancholy in which he argues that mourning arises from the loss of a significant other and melancholy from the loss of the self. Melancholy is associated with death, or at least with the (v)alley of the shadow of death (the precursor to death) and with mourning (the aftermath of death). Melancholy is the immanent counterpart to the transcendental sublime and the spiritual counterpart to the psychological uncanny. As the sublime and uncanny are a secular theology for a world in which God is dead, so melancholy is a secular spirituality in the lower psychopathological register. Melancholy is the neurosis symptomatic of capitalist society driven by greed for, and indulging in gluttony of, the earth's resources in the Anthropobscene, rather than exercising gratitude for her generosity in the desire for mutuality. This chapter also focusses on Walter Benjamin's work on the German mourning play, E. M. Cioran's discussions of despair that he associates with mud, Wilkie Collins's evocation of the black water of despair, and Charles Dickens's portrayals of "mudfog" and "marshy commons" in England and "the grim domains of giant despair" in an American swamp.

The mixing of the elements of earth and water occurred in the slime of World War I trench warfare. Chapter 7 of *Living with Nature in the Anthropobscene* brings western philosophy, ecoculture, and psychopolitical ecology together found literally in the mixing of the elements of earth and water in World War I trench warfare beginning with firsthand accounts of those who were there and wrote about it, focussing especially on mud and hell in the Anthropobscene. The transformation in both world wars of the natural and cultural environments in the country and the city and shifts in the conduct of industrialised world warfare are indicative of a change in the conduct of war at the fundamental level of the four elements of earth, air, fire, and water. In the case of the WWI, with its heavy artillery and trench warfare, earth and water were mixed to produce what commentators and caption writers called "swamps," while war photographers documented and dramatised them, in Frank Hurley's case the s(ub)lime in a single composite with a cloud burst of beneficent sunlight above the horror of the muddy landscape of trench warfare. The transformation of the air took also place in both wars with aerial weapons. Air became an area of conflict in the third dimension of height (rather than a front in the two dimensions of length and breadth on the land). This shift took place with the development and deployment of chemical weapons during the first year of WWI. These transformations had a dramatic and devastating physical and mental impact on those who experienced them directly and indirectly.

By contrast with World War I, the mixing of the elements of fire and air took place in the sublime bombing of World War II aerial warfare. Chapter 8 of *Living with Nature in the Anthropobscene* brings western philosophy, ecoculture, and psychopolitical ecology together in the mixing of the elements of fire and air in World War II aerial warfare beginning with the theory of war in the air in military texts and science fiction. It focusses on the aftermath of the terror of the ruinous landscape below of aerial bombing in the Anthropobscene. WWII was fought over (in two senses) the city; it was fought on and over the ground and sometimes in the air. The theatre of operations was through the countryside to the cities and over the cities. WWI was fought largely in the country and over the country; it was also fought on, in and sometimes under, the ground. The landscape of trench warfare in WWI was a battlefield for land-based military combatants of artillery and infantry pitted against each other in the country, whereas the landscape of fire-bombed cites in WWII was the result of airborne military operations of planes directed against civilians in the city, such as the fire-bombing of German cities, including Hamburg, a city set in a swamp. A first-hand, eye-witness recount of Hamburg emphasised the uncanny. Both world wars had an effect on the four elements and the six senses. This chapter concludes by considering the effects of both world wars on the four elements and the six senses in the Anthropobscene.

The mixing of the elements of earth, air, fire, and water occurred in the development of modern cities culminating in the Anthropobscene. Chapter 9 of *Living with Nature in the Anthropobscene* brings western philosophy, ecoculture, and psychopolitical ecology together in the sublime city and slimy slums focussing on London, another city set in a swamp, beginning with its recent writers and then its nineteenth-century ones. The lost and repressed slimy swamps return in a fascination with, and horror of, the slimy slums of the city figured and denigrated as jungle or swamp in some late nineteenth-century writers about London. These writers are placist by constructing a circularity of figurality. The pejorative associations of a jungle or swamp are applied to the feral city slum and the jungle or swamp comes to be seen in the same pejorative light (or dark) as the city slum. Beside(s), or above, the nether world of the uncanny city of dreadful night, nineteenth-century London also possessed and displayed the upper or fore world of the sublime city of delightful light encapsulated in the Crystal Palace built in London for the Great Exhibition of 1851 that fascinated Fyodor Dostoevsky as the quintessential symbol of modernity. The sublime city is made possible by draining and building on its slimy swamps and it makes slimy slums. All three go hand-in-hand.

The Greek philosophy of the four elements was syncretised with Judeo-Christian theology in the seventeenth century CE by John Milton in *Paradise Lost* following in Dante's footsteps. Chapter 10 of *Living with Nature in the Anthropobscene* brings western philosophy, ecoculture, and psycho- and femo-political ecology together in tombs, wombs, and being born beginning with Milton's rewriting in the seventeenth century of the biblical story of creation out of what he called "the womb of Nature," the pre-mixing of the four elements, an apt descriptor for fertile wetlands. In imperialist adventure-romances, such as H. Rider Haggard's *She* and Joseph Conrad's "The Lagoon," the journey of the hero into the swamp figures the horror of repeating birth in reverse of going up the birth canal. J. G. Ballard's post-apocalyptic dystopian novel, *The Drowned World*, is concerned with the return of the spatial repressed (the tropical colonies and the uterus), temporal repressed (the uterine and colonial pasts) and psychological repressed (memories of the womb and the mother's body). In Janet Frame's story of the same title as Conrad's, the lagoon represents a way for the first person narrator of establishing and celebrating matrilineal descent. Lagoons are some of the most fertile ecosystems and habitats on the planet. Yet the lagoon at low tide is not a place of new life, but of death and destruction. This is a part of the rhythm of life and death. Lagoons and other wetlands are places of what Hannah Arendt called natality, or being born, that she countered to Martin Heidegger's morbid solipsism of being towards death in the Anthropobscene.

Symbiosis, livelihood, and bioregion are ways of living and being with nature in the Anthropobscene and countering sublimation with desublimation in the Symbiocene. Chapter 11 of *Living with Nature in the Anthropobscene* brings western philosophy, ecoculture, and psycho- and femo-political ecology together in discussions of symbiosis, livelihood, and bioregion beginning with Lynn Margulis's work on bio-symbiosis, Margaret Mahler's work on psycho-symbiosis, Jessica Benjamin's work on bonding and Michel Serres's work on parasitism and the natural contract. It then focusses on Raymond Williams's late ecocultural work on livelihood and socialist ecology. It goes on to focus on the biogeographical concept of bioregion, the airshed and watershed where one lives. A bioregion always sustains lives and livelihoods in relationship with the living earth. This relationship is situated on a continuum between the mutually beneficial and 'normal' desire for and pleasures of biological and psychological symbiosis and mutuality at one end, and the parasitical and psychotic drive for mastery at the other. The final chapter and the book concludes by advocating for psycho- and femo-political symbiosis and mutuality with the living earth in the Symbiocene.

References

Albrecht, Glenn A. 2019. *Earth Emotions: New Words for a New World*. Ithaca, NY: Cornell University Press.

Bishop, Ryan, and Jussi Parikka. 2013. The Anthropobscene: The Elemental Media Condition. *Machinology: Machines, Ecology, and Some Media Theory*. Accessed April 23, 2025. https://jussiparikka.net/2013/12/05/the-anthropobscene-the-elemental-media-condition/.

Crutzen, Paul. 2006. The "Anthropocene". In *Earth System Science in the Anthropocene: Emerging Issues and Problems*, ed. Eckart Ehlers and Thomas Krafft, 13–18. Berlin, Heidelberg: Springer. Accessed April 23, 2025. https://doi.org/10.1007/3-540-26590-2_3.

Giblett, Rod. 2011. *People and Places of Nature and Culture*. Bristol: Intellect Books.

Giblett, Rod. 2018. Cultivating Resilience in the Anthropobscene: Resources for a Journey of Hope Towards the Symbiocene. *Annual Literature Program Lecture, Susquehanna University, April 8*. Accessed April 23, 2025. https://www.susqu.edu/live/news/452-april-events-at-susquehanna-university.

Giblett, Rod. 2025. *Waking to Nature with Thoreau and Benjamin: Or, Psychopolitical Ecology*. New York: Bloomsbury.

Giblett, Rod. 2026a. *The Politics of Nature and Environmental Writing: Political Ecocriticism*. New York: Palgrave Macmillan.

Giblett, Rod. 2026b. *Wise Women for Loving Nature: Or, Femopolitical Ecology*. New York: Bloomsbury.

Mumford, Lewis. 1934. *Technics and Civilization*. New York: Harcourt Brace Jovanovich.

Sloterdijk, Peter. 2025. *The Terrible Children of Modernity: An Antigenealogical Experiment*. Translated by Oliver Berghof. New York: Columbia University Press. First published in German in 2014.

Sofoulis, Zoë. 1988. Through the Lumen: Frankenstein and the Optics of Re-Origination. Ph.D. Thesis, History of Consciousness, University of California, Santa Cruz.

CHAPTER 2

The Elements, the Humors, and the Monstrous

Abstract The mixing of the qualities of hot, cold, wet, and dry create the four elements of earth, air, fire, and water for ancient Greek philosophy and ecoculture. The elements themselves should not mix, such as earth and water mixing to make wetlands. The mixing of the elements of air and water also gave rise to malaria, both literally bad air and a disease. The element of water gave rise to monsters; most monsters are water-beings, or mix water with one or other of the elements. The elements also gave rise to humors, or psychological states, even psychosomatic affects. The mixing of the qualities of cold and dry gave rise to the element of the earth, and in turn to the humor of melancholy, one of the persistent maladies of modern western culture in its transformations into sloughs of despondency and mires of depression. Melancholy is associated with mourning and death, so detective stories have been particularly fascinated with it. This chapter traces the genealogy of the elements and the humors, especially melancholy, and the monstrous in western philosophy, culture and medicine from Aristotle, Plato, Hippocrates, and Hercules through seventeenth-century philosophy and theology and nineteenth-century detective fiction to twentieth-century cultural studies.

Water is one of the four elements for ancient Greek philosophy and ecoculture. The other three elements for them are earth, fire, and air. The elements were created by mixing the qualities of hot, cold, wet, and dry. The

R. Giblett, *Living with Nature in the Anthropobscene*,
https://doi.org/10.1007/978-3-032-12822-5_2

elements in ancient Greek philosophy and ecoculture themselves should not mix, such as earth and water mixing to make wetlands. The mixing of the elements of air and water also gave rise to malaria, both literally bad air and a disease. The element of water gave rise to monsters; most monsters are water-beings, or mix water with one or other of the elements. The elements also gave rise to humors, or psychological states, even psychosomatic affects in the Anthropobscene. The mixing of the qualities of cold and dry gave rise to the element of the earth, and in turn to the humor of melancholy, one of the persistent maladies of modern western culture in the Anthropobscene, including in its transformations into despondency and depression. Melancholy and despondency have been the happy hunting ground for writers from the seventeenth century, especially in sloughs. Melancholy is associated with mourning and death, so detective stories have been particularly fascinated with it, especially in mires.

2.1 The Elements

The theory of the elements was developed by the pre-Socratic philosophers and later enunciated by Aristotle in the treatise which came to be called 'De Generatione et Corruptione' ('On Coming-to-be and Passing-Away'). For Aristotle (cited by Ross 1995, 104) what he called the four "so-called elements" of earth, air, fire, and water were produced by the selective "coupling" of the four "elementary qualities" and two "contrarities" of hot and cold, dry and wet (or "moist" as the translator prefers, or "fluid" as one commentator prefers). As a result earth is cold and dry, air is hot and moist, fire is hot and dry, and water is cold and moist. Theoretically it was possible for the four "elementary qualities" and two "contrarities" of hot, cold, dry, and wet to be combined into six couples, but as it is impossible for something to be both hot and cold, wet and dry (though not in practice for a solid to be damp or sodden as in a slimy wetland, nor for a solid to be sublim(at)ed into a gas by heat), there were in effect only four proper couples. It was thus acceptable for the elementary and contrary qualities to mix, but only in four prescribed or proper ways to produce the four elements. The mixing of the "contrarieties" of hot and cold, wet and dry was "impossible," or in David Ross's (1995, 105) terms they "refuse to be coupled," though Aristotle (1930, §§329a–332b) later concedes that "all the 'elements' will be able to be transformed out of one another" which would involve coupling of the contrarities.

Wetlands were effectively ruled out of court in this cosmogony on two counts depending on whether they were considered at the level of elements or qualities. They were ruled out at the level of elements because they combine earth and water, or because they mediate between earth and water spatially or temporally. Wetlands are in-between earth and water. They were also ruled out at the level of elementary qualities for as earth is cold and dry, and water cold and wet, a mixture of or a mediating category between the two sets of contrarieties would produce the quality of cold, dry, and wet, an evident impossibility in this philosophical schema. The sublime is even more out of court as it involves a transition through the elements from 'earth' (or solid) to 'air' (or gas) by or via 'fire' (or heat). It would also entail the absurd combination of the qualities of hot, cold, dry, and wet.

Aristotle distinguishes the elementary qualities of the hot and cold from the wet and dry in terms of the active/passive distinction: "the first pair implies *power to act* and the second pair *susceptibility*" (§329b). Or as David Ross (1995, 105) put it, "hot-cold play in general the part of agent and dry-fluid that of patient." The distinction between agent and patient is gendered in patriarchy. The former is associated with virile masculine activity and the latter with supine feminine passivity. Aristotelian philosophy is no exception as it for Carolyn Merchant (1980, 13) "associated activity with maleness and passivity with femaleness." On the Aristotelian view of (re)production, the masculine hot-cold inseminates the feminine wet-dry and brings forth the four elements. A gender distinction also applies to the elements. Of the four elements for Aristotle "Fire and Earth [...] are extremes and purest" and "Water and Air [...] are intermediaries and more like blends" (§§330b–331a). In other words, fire and earth are definite and discrete, and 'masculine' in patriarchal terms; water and air are mediatory and mixed, and 'feminine' in patriarchal terms. Fire and earth are privileged over air and water, hot and cold over wet and dry so the table below (Table 2.1) is hierarchical.

Yet Aristotle accepted that "Earth has no power of cohesion without the moist" (presumably just as long as the moist stayed between the bits of earth and did not mix with them thus dampening them) (§335a). 'Feminine' moist (the wetlands), holds 'masculine' earth (the dry lands) together; without the former the latter would fall apart. Wetlands are, as an article in *National Geographic* argues, "the glue that holds the land together" (Mitchell 1992, 15). More precisely, wetlands are the soft glue that holds, in the words of Liam Davison's *Soundings*, "the hard bones of

Table 2.1 The four qualities, elements, and humors

	Hot	*Cold*
Dry	Fire	Earth
	Heat	Solid
	Choler/ic	Melancholy/ic
'Wet'	Air	Water
	Gaseous	Liquid/Fluid
	Blood/Sanguinous	Phlegm/Atic

the land" together (Davison 1993, 85). The moist, moreover for Aristotle, is "that which being readily adaptable in shape is not determined by any limit of its own" (§329b). This 'feminine' adaptability and limitlessness of water is the condition of possibility for modern 'masculine' waterworks.

Yet while Aristotle was appreciative of the cohesive and other qualities of the moist, he certainly did not like the inundatory or penetrative qualities of the damp and the sodden which result from the solid and the liquid, earth and water, the dry and the wet, the masculine and the feminine "impossibly" combining in aberrant ways: '"damp' is that which has foreign moisture on its surface ('sodden' being that which is penetrated to its core)" (§330b). In Aristotelian terms a wetland would be earth with "foreign" moisture on it, as in a lagoon, or earth "penetrated" with moisture to its core, as in a marsh. Either way, 'masculine' earth should not have 'feminine' water lying on its surface, nor penetrating its depths. The only proper function for 'feminine' water is to make the 'masculine' earth stick together. On this Aristotelian view and by implication, earth should not have the "foreign" matter of water on it, nor should it be "penetrated" by the invasive power of water. Earth should be dry for Aristotle and not wet.

2.2 Unhealthy Waters

The mixing of the elements of air and water gave rise not only to wetlands, but also to malaria, both literally bad air and a disease. Probably the earliest extant discussion of the association between wetlands and disease in western culture is in the Hippocratic writings. In the essay 'Airs, Waters, Places,' the Hippocratic writer is concerned to distinguish "the effect of different kinds of water, to indicate which are healthy and which unhealthy, and what effects, both good and bad, they may be expected to produce" (Lloyd 1978, 152). Within the category of unhealthy waters are:

> stagnant waters from marshes and lakes [which] will necessarily be warm, thick and of an unpleasant smell in summer. Because such water is still and fed by rains, it is evaporated by the hot sun. Thus it is coloured, harmful and productive of biliousness. In winter it will be cold, icy and muddied by melting snow and ice. This makes it productive of phlegm and hoarseness. (Lloyd 1978, 152)

The writer goes on to enumerate the features of the physiology and physiognomy of swamp dwellers.

The Loeb Classical Library translation of this essay also elaborates the description of the waters in the following miasmatic terms:

> turbid, stagnant waters in marshes and swamps are hot, thick and evil-smelling in summer because of their stagnation and failure to flow [...] Waters that issue level on the face of the earth and such as are marshy, standing and stagnant must in summer be hot, thick and stinking, because there is no outflow. (Cited in Mattock and Lyons 1969, 48)

Not only does the turbidity of swamp waters mean that they mediate between solidity and liquidity (and so violate these categories), but their stagnancy and hotness violates the qualities of fluidity and coolness assigned to waters.

Waters, for the Hippocratic writer, should flow and be cool; waters which do not flow and are not cool violate the order of things. Hot, stagnant, and turbid waters are also evil-smelling, whereas cool, flowing, and clear waters are pleasant-smelling. Across a range of categories and qualities in the Hippocratic taxonomy of waters, wetlands are ruled out of court. On top of all this, and because of all this, they are bad for bodily health. For what reasons, the writer does not say, but there is a certain obviousness that this degree of category- and quality-mediation and violation is such a disruption to the order of things that it must be detrimental to health.

Such a view persisted for two millennia until the discovery in the 1890s of the anopheles mosquito as the vector for malaria (Giblett, 1996, chapter 5). The denigration of wetlands, especially in relation to health, was exported to, or imported into, England and was taken up by its writers in relation to Fens and other wetlands. For instance, in the early seventeenth century of the common-era Shakespeare's *King Lear*, Lear's ultimate repudiation of Cordelia is couched in the Hippocratic and miasmatic terms of "Infect her beauty,/ You fen-sucked fogs." In Shakespeare's *The Tempest*, Caliban curses Prospero in similar terms: "All the infections that the sun sucks up/ From bogs, fens, flats, on Prosper fall, and make him/ By inchmeal a disease." Fens are employed in both instances as a source of insult against the health and beauty or life of someone who has been repudiated or is reviled. On the basis of the miasmatic theory of malaria the Sanitary Movement of two centuries later was founded.[1]

2.3 The Humors

In the western theory of the elements, the four elements combine to form or are associated with psychosomatic states or 'humors' as developed by the Elizabethans and as formulated by E. M. W. Tillyard (1963, 76) in his classic study of the Elizabethan world picture. These states can be tabulated in terms of the mixing of the four elementary qualities into the four legitimate elements, in turn into four more modern qualitative elements and finally into four proper substances and humours (see Table 2.1)

This schema becomes problematic when considering the association between melancholia and wetlands as in the terms of this schema they should not strictly be associated. Or more precisely, wetlands as between, or combining, earth and water should be seen as equally phlegmatic as they should be seen as melancholic. As a result this combination of elements would produce a phlegmatic melancholy, an aberrant humour.

For Aristotle the arrangement of the elementary qualities and their proper couples were associated with physical conditions and with mental humors or temperament. In his treatise 'De Problemata' in discussing "problems connected with the effect of locality on temperament," Aristotle goes on to answer with another question the question he poses:

> why is it that those who live in airy regions grow old slowly, but those who inhabit hollow and marshy districts age quickly? Is it because old age is a process of putrefaction, and that which is at rest putrefies, but that which is in motion is either quite free from, or at any rate less liable to, putrefaction, as we see in water? In lofty regions, therefore, owing to the free access of the breezes, the air is in motion, but in hollow districts it stagnates. Furthermore, in the former, owing to its movement, the air is always pure and constantly renewed, but in marshy districts it is stagnant. (Aristotle 1927, §909b)

[1] For further discussion of the Sanitary Movement see Giblett (2024, chapter 4).

Not only is the water stagnant in marshy districts, but the air as well. No wonder such districts are bad for you both physically and mentally!

Despite the differences between his work and that of his teacher, he reproduces, like a good pupil, Plato's denigration of "hollow places" as Socrates undertook in the *Phaedo*:

> there are many hollow places all around the earth, places of every shape and size, into which the water and mist and air have collected. But the earth itself is as pure as the starry heaven in which it lies, and which is called Ether by most of our authorities. The water, mist, and air are the dregs of Ether, and they are continually draining into the hollow places of the earth. We do not realise that we are living in its hollows, but assume that we are living on the earth's surface [...] this earth and its stones and all the regions in which we live are marred and corroded by the brine, and there is no vegetation worth mentioning, and scarcely any degree of perfect formation, but only caverns and sand and measureless mud, and tracts of slime wherever there is earth as well; and nothing is in the least worthy to be judged beautiful by our standards. But the things above excel those of our world to a degree far greater still. (Plato 1954, §109a–110d)

Socrates valorises the extraterrestrial and the transcendental over the immanent—the intraterrestrial and the aquaterrestrial. Rather than scenery, wetlands are obscenery.

The creation of this Platonic split and hierarchy in the *Phaedo* is the primal birth of western philosophy insofar as the concept of nature is concerned. More precisely, it gave birth to the upper stratum of the (idealist) sublime, and valorised it over the lower stratum of (materialist) slime (as we will see with both the sublime and slime in Chap. 3). It made possible the birth of the sublime culminating in modernity and its industrial capitalist technologies (as we will also see in Chap. 3). It also created the conditions for the birth of the Anthropobscene.

The birth of western philosophy is coincidental with what Peter Sloterdijk (2024, 91) calls "the primal scene of 'occidental' philosophy" of the death of Socrates related in Plato's *Phaedo* (§117 a–b; cited by Sloterdijk 2024, 39). Sloterdijk (2024, 39) calls the death of Socrates "the Passion of Europe's archphilosopher," akin to the Passion of Jesus, Europe's archtheologian of Christianity. Plato's *Phaedo* for Sloterdijk (2024, 91) is one of "the initiatory texts of philosophy par excellence." The Socratic biographer and dialogist Plato is the founder of western philosopher, while his Christian counterpart, the apostolic, epistolic, and travelling evangelist Paul, is "the founder of the Christian religion" for Sloterdijk (2020, 130). Sloterdijk (2024, 108) also calls the *Phaedo* "this sublimest dialogue," though not only in the sense he means. It is sublime, not only for the death of Socrates, but also for the birth of the sublime and the sublimation of slime. Sloterdijk does not discuss the passage of §109a–110d from *Pheado cited above.* He is concerned here with philosophers, the authors of texts, rather than with philosophy and critiquing the sublime and sublimation.

Socrates for Sloterdijk (2012, 29) was "on an interior journey" and "the inventor of a sublime emigration." Socrates was indeed on an interior and exterior journey into the sublime on three pathways: the invention of philosophic self-sublimation into death; textual sublimation into immortality via the vehicle and vector of Plato's written dialogues; and sublimation of slime into the sublime before the terms were invented. "The philosopher [is] the sublimator" as Sloterdijk (2024, 38) says, applicable not only to Socrates, but also to himself when he sublimates with the fiery heat of his enflamed rhetoric solid matter, liquid water, and gaseous air into bubbles, foams, globes, and spheres. The philosopher is also the sublator, such as when Sloterdijk (2025, xlvi, 210) invokes sublation (*Aufhebung*), the third and final process/phase in the Hegelian dialectic, the antithesis of the antithesis, aka synthesis.

Psychopolitical ecology counters the sublimating project of western philosophy. In particular, it desublimates the public and/or private spheres of the bubbles of microspheres in Sloterdijk's terms in the first volume of his 'Spheres' trilogy (Sloterdijk 2011), the globes of macrospheres in the second volume of his 'Spheres' trilogy (Sloterdijk 2014) and the foams of plural spheres in the third volume of his 'Spheres' trilogy (Sloterdijk 2016). Psychopolitical ecology desublimates these spheres into the home habitats of bioregional places, the biosphere of living beings, and the ecosphere of the earthly household (as we will see in later chapters). The psychopolitical ecologist is the desublimator.

The death of the father of western philosophy killed by his sons ("the primal scene") was also the primal birth of philosophy insofar as the concept of culture is concerned. More precisely it gave birth to what Sloterdijk (2024, 90) calls "the metaphysical death-appetites" culminating in Heidegger's "will to death," Fascism's "vitalism of the dead" as Sloterdijk (1987, 286) calls it, and Sartre's misogynist horror of slime (as we will see in Chap. 3). The deaths of the fathers of ancient and modern philosophy and Christian theology gave birth to all of them. A strange, spectral, resurrected new life came out of death culminating in Heidegger's morbid solipsism of "being towards death" countered by his student's Hannah Arendt's natality of "being born," and the biodiversity of slimy swamps (both discussed in Chap. 10). Sloterdijk valorises the biographies of the great philosophers and their giving birth to philosophy. He overlooks the birth of culture and nature in philosophy, and its denigration of wetlands. The *Phaedo* is the slimiest dialogue. It perfectly illustrates Zoë Sofoulis's thesis that "slime is the secret of the sublime" that she encapsulates in the parenthetical portmanteau "s(ub)lime." The sublime is sublimated out of slime and threatens desublimation back into it (as we will see in Chap. 3).

Is not Sloterdijk on the threshold of philosophy with the door wide open moving in and out through it? He is willing to venture far and wide outside the constricting confines of philosophy after shaking off the shackles of Adorno and Horkheimer, but is he not also reluctant to relinquish his consoling connection with it, especially Heidegger who he calls "the greatest thinker of Old Europe" (in the material and genitive, including possessive, senses) and "the avant-garde conservative," an oxymoron (Sloterdijk 2016, 481, 674)? Is not Sloterdijk also a little dismissive of Walter Benjamin as a philosopher, the greatest thinker for New Europe and the avant-garde radical? Is not Sloterdijk also seduced by the siren song of the music of Habermas's public sphere (despite their spat) when he sublimates with the fiery heat of his enflamed rhetoric solid matter, liquid water, and gaseous air, the biosphere and the ecosphere, into the bubbles, foams, and globes of various forms of the public and private spheres?[2] Is not Sloterdijk the student, the logos-ian, of the microspheres, macrospheres, and plural spheres?

2.4 Monsters and the Monstrous

The wetland is also the home of monsters. In the pre-Socratic cosmogony, at least in Christian Athenagoras's second-century account of it, "water was the origin for the totality of things [...] and from water slime was established, and from both of them was generated a living creature, a snake with a lion's head growing on to it," the king of anti-beasts, or monsters (Kirk et al. 1983, 25). The sixth labour of the heroic Heracles (Hercules) in Greek mythology is to kill the Stymphalian birds, "the

[2] For critical discussion of the public and private spheres within the biosphere and the ecosphere drawing on Habermas's and Hannah Arendt's work, see Giblett (2011, chapter 2).

countless brazen-beaked, brazen-clawed, brazen-winged, man-eating birds, sacred to Ares" which "had flocked to the Stymphalian Marsh" (Graves 1955, 119). They are waterbirds and war-birds—wa(te)rbirds for short! From the marsh the wa(te)rbirds occasionally took to the air in "great flocks to kill men and beasts by discharging a shower of brazen feathers and at the same time muting a poisonous excrement which blighted the crops" (Graves 1955, 119). No doubt this poisonous excrement is produced by the birds because they live in and feed greedily on the ostensibly decaying and rotting life of the marsh. In short, they are coprophagic.

The birds represent the antithesis of, or perhaps more precisely an offence (in at least two senses) towards, agriculture. In response, even resistance, to this myth my reading is a defence of the marsh and its birds. The marsh and the martial are allied in the myth against agriculture and men (andros). The ostensibly martial marsh birds are figured as misandropic in this foundational myth of patriarchal western culture which displaces the blame onto the victim for the war men have fought against the marsh no doubt from before heroic Heracles's time. This martial anti-marsh campaign culminates in Fascist Mussolini's boast that the war against the Pontine Marshes was the kind of war he liked to fight, and win, at least in the short term.

Instead of men (and agriculture) being shown as antagonistic to the marsh, the marsh is made out to be monstrously martial towards men (and agriculture). A classic case of blaming the victim! The martial, men and agriculture (and the matrix) have been allied against the marsh, the maternal and marsh-birds in mainstream western culture. Marsh and agriculture are opposed but they can (or should) co-exist, whereas the Stymphalian birds are made to represent a martial/marsh-al oral and anal sadism which would destroy the crops of the latter and vanquish the former over the latter. The oral and anal sadism which men have exercised against the marsh by eating it up, grazing their animals on it (consuming its resources), administering an enema to it (draining it) and crapping in it ('sanitary land-filling' it) is here displaced onto anthropophagic, or more precisely androphagic, coprophagic and coprophilic waterbirds. As they feed on shit and shit shit, their shit is not manure, but poison.

Heracles is initially unable to drive the birds away with his arrows. Typically, even archetypically, he finds that "the marsh seemed neither solid enough to support a man walking, nor liquid enough for the use of a boat" (Graves 1955, 119). The wetland for Heracles, and for hundreds of generations of heroic westerners since him, is neither solid land nor deep water but mediates (between) the two. The wetland is centrally construed in mainstream western culture as a problem of transportation as it is for Allnutt in C. S. Forester's *African Queen* whose narrator remarks of the Swamp of the Bora that for Allnutt and Rose "no place could be beautiful that presented navigational difficulties," especially for the hero on a mission or a quest or with a job to do which requires him to traverse the place with some end or goal in view or mind (Forester 1956, 135). Yet rather than a hero on a single-minded mission to conquer the marsh by crossing it (though in one version of the myth Heracles drained the marsh (Graves 1955, 120)), Heracles's labor is to kill the horrific wa(te)rfoul/wa(te)rfowl.

This "labor" "glorifies" Heracles, according to Robert Graves, as:

> the healer who expels fever demons, identified with marsh-birds. The helmeted birds shown on Stymphalian coins are spoon-bills, cousins to the cranes which appear in English medieval carvings as sucking the breath of sick men. They are, in fact, bird-legged Sirens, personifications of fever [...] The spoon-bill is closely related to the ibis, another marsh-bird, sacred to the god Thoth, inventor of writing. (Graves 1955, 120–121)

The Stymphalian birds personify, or more precisely 'femininify' as they were Sirens and women according to some accounts (Graves 1955, 120), the miasmatic diseases to which the marsh gives rise. How spoonbills, a benign species of waterbird which feeds on crustaceans and small fish by sweeping its spoon-shaped bill from side to side under the water, came to femininify disease and to be blamed for the war men have fought against the marsh is not only a mystery, but also a patent injustice (Kingsford 1991, 70).

A limited explanation of this mystery and injustice is offered by Gaston Bachelard (1983, 101) when he suggests that the marsh in this myth becomes what he calls:

> 'stymphalized.' It becomes the black swamp where live monstrous birds, the Stymphalides, 'nurslings of Ares, which shoot their feathers like arrows, which ravage and soil the fruits of the earth, which feed on human flesh.' This stymphalization [...] corresponds to a particular feature of melancholy imagination.

I suppose only a deeply melancholic, even manic depressive, imagination could see a spoonbill as a bird which wars on humans, destroys their crops by excreting on them and eats their flesh, as, in short, both anally and orally sadistic! As a result of such a diet it is hardly surprising then that they "mute" poisonous excrement! Such excrement is thus the result of their diet of human crops and flesh rather than the product of the marsh in which they live. Hercules (Heracles) for Sloterdijk (2013, 194) is "the patron saint of athletes." Hercules is also the patron saint of bloodsport psychozoopaths who kill waterbirds for fun.

Wetlands are considered to be dangerous to life from disease and can lead to death. They are even the land of the dead. Hades, the land of the dead, is a wetland for Homer. In *The Iliad* "misty Tartaros" is "the deepest gulf below earth" and in Hesiod's *Theogony* it is "the grievous, dank limits which even the gods detest" (cited by Kirk et al. 1983, 9 and 40). Similarly in Homer's *Odyssey*:

> [at…] the limit […]of the deep-running Ocean
> There lie the community and city of Kimmerian people,
> hidden in fog and cloud, nor does Helios, the radiant sun, ever break through the dark to illuminate them with his shining,
> neither when he climbs up into the starry heaven
> nor when he wheels to return again from heaven to earth,
> but always a glum night is spread over wretched mortals. (Homer 1965, XI 12–19, 168)

The ancient Romans took on and developed this view (Giblett 2024, chapter 5), as did John Milton, the seventeenth-century syncretist of Judeo-Christian theology and Greek philosophy (as we will see shortly and in Chap. 10).

Marsh monsters are the product not only of a melancholic imagination but also, and in turn, of a miasmatic theory of monstrosity. Not only does the marsh 'cause' physical and 'mental' illness due to the vapours which rise from it, but also the miasma give 'birth' to monsters. In the Old English elegiac long poem *Beowulf* Hrothgar describes Grendel the marsh monster and his monstrous mother in miasmatic terms as two:

> huge wanderers of the marches guarding the moors, alien spirits […] They dwell in a land unknown, wolf-haunted slopes, wind-swept headlands, perilous marsh-paths, where the mountain stream goes down under the mists of the cliff,—a flood under the earth. It is not far hence, in miles, that the lake

> stands over which hang groves covered with frost: the wood, firm-rooted, over-shadows the water. There may be seen each night a fearful wonder, [...] fire on the flood! [...] That is no pleasant spot. Thence rises up the surging waters darkly to the clouds, when the wind stirs up baleful storms, until the air grows misty, the heavens weep. (*Beowulf* 1950, 89, 90)[3]

The wetland not only mixes the elements of water and earth, and air and water when the atmosphere is misty in the temperate zone (and dripping wet in the tropics), but also fire and water when marsh gases are ignited in the temperate zone. The traditional dragon breathing fire is arguably just a marsh (monster) 'breathing' ignited marsh gases. Fire is also associated with hell, with the burning fires of brimstone, so the wetland in patriarchal western culture is a kind of watery hell.

Grendel and his mother are alien spirits because they are "Cain's kindred" (*Beowulf* 1950, 25) condemned by God to the wetland hell guilty, presumably like Cain, of fratricide though who their brother is not specified. The worst of the two is "that grim spirit [...] called Grendel, the renowned traverser of the marches, who held the moors, the fen and fastness; unblessed creature, he dwelt for a while in the lair of monsters, after the Creator had condemned them" (25). Grendel, however, has not only been condemned but also been made into the personification, or more precisely 'monstrification,' of damnation. He is described as "the grim and greedy creature of damnation, fierce and furious," and as "the demon, the dark death-shadow" who "in the endless night [...] held the misty moors" (26 and 28). Grendel is a demonic agent, or 'angel,' of death who lives in the wetland realm of darkness.

Grendel's mother is not much better as she is described as "monstrous among womenkind [...] who must needs inhabit the dread waters, chilling stream" (84). Grendel's mother lives in the dreadful aquifer, the flood beneath the earth, which is even worse than the flood on or over the earth. Grendel's mother is an instance of what Barbara Creed (1986) calls "the monstrous-feminine," or more precisely the monstrous-maternal, a patriarchal figuring of the mother as monster. Grendel and his mother are both specifically marsh monsters and a part of a long genealogy.

[3] Compare a similar, later Medieval wetlandscape: "the heavens are lifted high, but under them evilly/ mist hangs on the moor, melts on the mountain;" *Sir Gawain and the Green Knight* [...] (1975, 77).

Grendel is no doubt a prototype for Satan in John Milton's epic long poem *Paradise Lost* first published in 1667. Milton's Satan is a swamp serpent and marsh monster, or more precisely a miasmatic marsh monster. He is described as "involved in rising mist" (IX, 75) and as "wrapped in mist/Of midnight vapour" who glides "obscure" (IX, 158–159) "like a black mist low creeping" (IX, 180). Satan is a miasmatic marsh monster because as a fallen angel he has been "mixed with bestial slime,/This essence to incarnate and imbrute" (IX, 165–166). Satan 'personifies,' or more precisely 'monstrifies,' slime and wears as his proper wardrobe the miasma which arise from it. He monstrifies slime because he was "engendered in the Pythian vale on slime" (X, 530). He is born from slime and made of slime; he is slime by birth and slime by nature.

For Milton it seems also as Stephen Greenblatt (1982, 12) argues for Martin Luther, "the Devil dwells in excrement." As a slimy swamp serpent his mode of propulsion is not restricted to any one kind, but varies according to the (wet)landscape over and through which he is travelling. As he does not stick to one mode of propulsion he is aberrant: "so eagerly the Fiend/ O'er bog or steep, through strait, rough, dense, or rare,/With head, hands, wings, or feet pursues his way,/And swims or sinks, or wades, or creeps, or flies" (II, 947–950). Satan is even a flying miasmatic marsh monster and swamp serpent. He is monstrous because he combines the physical characteristics of birds (he flies (like the Stymphalian birds)), fish (he swims), reptiles (he creeps), and quadrupeds or bipeds (he sinks or wades).

Satan, in a word, swarms, and hence comes under the Biblical interdiction that "every swarming thing [...] is an abomination" (Leviticus 11:41). Swarming creatures, Mary Douglas (1966, 56) has commented, are:

> both those which teem in the waters and those which swarm on the ground. Whether we call it teeming, trailing, creeping or swarming, it is an interminable form of movement [like slime] [...] 'swarming' which is not a mode of propulsion proper to any particular element, cuts across the basic classification [like the wetland]. Swarming things are neither fish, flesh nor fowl. Eels and worms inhabit water, though not as fish; reptiles go on dry land, though not as quadrupeds; some insects fly, though not as birds. There is no order in them [...] As fish belong in the sea so worms belong in the realm of the grave with death and chaos.

Milton's Satan is a worm which belongs in the tomb, rather than womb, of the swamp (as we will see in Chap. 10). By swarming in the elements of earth, air, fire, and water Satan is an abominable monstrosity who upsets the order of things.

2.5 Melancholy

Even though the melancholic has traditionally been associated with the cold and dry, a visit to the cold and wet/dry marshes of a wetland by the drylander can induce melancholy, whereas the indigenous wetlander by temperament is usually phlegmatic, but tends towards the melancholic. By temperament, the wetlander at home in the wetland is phlegmatic (associated with the watery), whereas the drylander visiting the wetland tends to be infected, or more precisely affected, by the melancholic (the earthy, though more precisely the earthy in combination with the watery).

One of the "causes of melancholy" for Robert Burton writing in his *The Anatomy of Melancholy* published in 1621 was "standing waters." He went on to elaborate in Hippocratic terms that:

> standing waters, thick and ill-coloured, such as come forth of pools and moat where hemp is steeped or slimy fishes live, are most unwholesome, putrefied, and full of mites, creepers, slimy, muddy, unclean, corrupt, impure, by reason of the sun's heat and still standing; they cause foul distemperatures in the body and mind of man, are unfit to make drink of, to dress meat with, or to be used inwardly or outwardly [...] So that they that use filthy, standing, ill-coloured, thick, muddy water, must needs have muddy, ill-coloured, impure and infirm bodies. And because the body works upon the mind, they shall have grosser understandings, dull, foggy, melancholy spirits, and be readily subject to all manner of infirmities. (Burton 1932, 224–225)

Although Burton enunciates a psychosomatic theory of the body and mind (after all, he is pre-Descartes and Cartesian dualism), he subscribes to a simplistic analogy, or even correspondence theory, based primarily on colour, between the natural environment and the human body. And although he does not subscribe to the miasmatic theory of physical illness, it is the standing waters themselves, rather than the vapours which rise from them, which are to blame for illness, both 'mental' and physical.

Not only can standing waters cause melancholy, but also the cold and dry air which rises from them. When Burton discusses "bad air" the worst air of all out of hot air, cold air, and cold and dry air is the last which he describes as:

> a thick, cloudy, misty, foggy air, or such as come from fens, moorish grounds, lakes, muckhills, draughts, sinks, where any carcasses or carrion lies, or from whence any stinking fulsome smell comes: [...] such air is unwholesome, and engenders melancholy, plagues and what not. (Burton 1932, 239)

Yet such air is surely not dry but wet. Under the traditional western theory of the elements and humors as formulated systematically by the Elizabethans, the wetland did not fit easily into the schema as we have already seen. In fact, melancholy was strictly associated with the dry upland of the agricultural country and the modern city, and even with capitalism as Terry Eagleton (1986, 41) argues that "melancholia is an appropriate neurosis for a profit-based society" and with the sublime as Kant (1960, 47; see also 1952, 121) postulates that "the feeling of the sublime is sometimes accompanied with a certain dread or melancholy."

In response to the modern melancholic city sublime, there was a massive attempt to displace blame onto the wetland for its physical and mental ills. Wetlands are what Lyotard (1991) calls "scapelands," the geographical equivalent of the scapegoat on which communal sins are heaped. Wetlands are more "scapeland" than landscape as evidenced by landscape painting. Lyotard (1991, 184) argues that "deserts, mountains and plains, ruins, oceans and skies enjoy a privileged status in landscape painting." Wetlands do not enjoy such a privileged status in landscape painting. The scapewetland is driven out to die in the wilderness and so the sins of the community are expiated, the community cleansed of its moral dirt by constituting the wetland as dirt, as "matter out of place," to use Mary Douglas's (1966, 35) terms. Wetlands are what Charles Cotton in the seventeenth century called "nature's shames and ills" (Fowler 1994, 375).[4] Wetlands are obscenery, as they were for Socrates.

The Elizabethan "world picture" as enunciated by E. M. W. Tillyard was not flattering to the modern melancholic and sublimated city-dwelling drylanders; it was more flattering to the pre-modern phlegmatic,

[4] For further discussion of the aesthetics of the wetlandscape in general, see Giblett (1996, chapter 1), and of the English Fens in particular, see Giblett (2024, chapter 6).

'slimated' and pre-sublimated fen-dwelling wetlanders. As wetlands did not 'fit the picture,' rather than changing the picture, the wetland had to be changed, or more precisely in Burton's case the dwellers in wetlands, including in cities built on or near wetlands (invariably 'foreign'), denigrated.[5]

After enumerating a list of places where the air is bad, Burton (1932, 240) concludes by asking "how can they be excused that have a delicious seat, a pleasant air, and all that nature can afford, and yet through their own nastiness and sluttishness, immund and sordid manner of life, suffer their air to putrefy, and themselves to be choked up?" The wetland, its standing waters, its bad (but wet) airs, and its filthy dwellers, either indigenous wetlanders or city-in-wetland dwellers, is the cause of melancholy displaced from the cold and dry uplands (and discussed in Chap. 6).

2.6 The Slough of Despond

Following in Burton's swampy footsteps half a century later, the most famous depiction in English culture of a depressed place (in two senses) is John Bunyan's "Slough of Despond" in *Pilgrim's Progress*, first published in 1678. Louis Marin (cited by Fritzell 1978, 528) described *Pilgrim's Progress* as "the most influential religious book ever composed in the English language." It is also the second most published book in the English language after the Bible and before the complete works of Shakespeare. No doubt part of its pre-eminent influence has been to educate generations of readers that a wetland functioning ecologically is not what it seems, and is, but an allegorical emblem for a sump of iniquity that would drag the unsuspecting and unwary Christian down and entrap him/her for all eternity. Bunyan (2008, 18) relates how "this miry slough is such a place as cannot be mended; it is the descent whither the scum and filth that attends conviction for sin doth continually run, and therefore it is called the Slough of Despond." Bunyan's theologized slough of despond is singing from the song sheet of the melancholic swamp composed by Robert Burton in his miasmatic *Anatomy of Melancholy.* He is also unwittingly singing from Socrates's song sheet of Platonic idealism in the *Phaedo.* He syncretised Christian theology with Greek philosophy, like John Milton, his near contemporary, in *Paradise Lost* first published in

[5] For further discussion of cities and wetlands, see Giblett (2016).

1667 a decade earlier (as have seen in this chapter and we will see again in Chap. 10).

The increasing secularisation of English culture meant that after Bunyan the slough of despond was divested of its Christian overtones, but not of its denigratory, misaquaterrist associations. Instead of being a place of evil construed in religious terms, it became a place of melancholia, a kind of secular despondency leading to despair. Charles Dickens, for example, has "the Slough of Despond" in *Hard Times*, as well as "the slough of inanity." "The social swamp" as Thomas Huxley (1989, 335) called "*la misère*" in 1888 could be "a Slough of Despond." Dickens saves his most scathing attack on the swamp as "the grim domains of Giant Despair" and melancholy for Chapter 23 of *Martin Chuzzlewit* (as we will see in Chap. 6).

The slough of despond, rather than being rehabilitated or its denigratory connotations reversed in the process of increasing secularisation, came to stand for the dark side of the Enlightenment and Romanticism despite their differences. The Slough of Despond, Peter Fritzell has argued:

> is the well-remembered antithesis to enlightenment and romantic thought [...] With the declaration of man's [*sic*] perfectibility and the affirmation of sublime and picturesque vistas, with the denial of man's imperfectibility and the negation of landscapes which do not fit the conventions of the sublime or the picturesque vistas of the nineteenth century can proceed to their new-found manifest destiny. They can proceed with the story of exploitation, the story of draining, ditching, clearing and filling, which will improve, civilise, humanise, and finally redeem the nonhuman environment. (Fritzell 1978, 529)

Such a place as the slough of despond as the antithesis of enlightenment would seem to make the ideal setting or backdrop for the private detective story as it is a place of almost impenetrable evil and darkness which can only be pierced by the penetrating light of reason brought to bear upon it by the superior intellect and insight of the great man himself (for it is invariably a man so why not a feminist detective story which inverts, or better still subverts, this scenario?).

It is hardly surprising then that the most famous private detective of them all in probably his most famous story of all should have to deal with a slough of despond. This is, of course, Sherlock Holmes in Arthur Conan Doyle's *The Hound of the Baskervilles* with its Great Grimpen Mire. When Stapleton takes Dr Watson on a tour of the moor, he asks Watson if he notices "'those bright green spots scattered thickly over it.'" Watson replies, "'yes, they seem more fertile than the rest'" to which his guide laughs and goes on to say "'that is the great Grimpen Mire'" which he describes as "'the bog-hole'" and as "'an awful place.'" By acknowledging the fertility of the mire, Watson was a wetlands ecologist before his time! Whilst the Mire is being discussed "a long, low moan, indescribably sad swept over the moor. It filled the whole air, and yet it was impossible to say whence it came. From a dull murmur it swelled into a deep roar, and then sank back into a melancholy, throbbing murmur once again." Watson asks what the sound is only to be told that "'the peasants say it is the Hound of the Baskervilles calling for its prey.'"

Watson is not satisfied with this answer, dismissing it as "'nonsense'" and pressing Stapleton for a more rational explanation which he feebly offers:

> 'bogs make queer noises sometimes. It's the mud settling, or the water rising, or something.' 'No, no that was a living voice.' 'Well, perhaps it was. Did you ever hear a bittern booming?' 'No, I never did.' 'It's a very rare bird – practically extinct – in England now, but all things are possible upon the moor …' 'It's the weirdest, strangest thing that ever I heard in my life.' 'Yes, it's rather an uncanny place altogether.'

Indeed, the wetland is the uncanny place *par excellence* for western culture (as we will see in Chap. 5).

Watson is greatly affected by "the melancholy of the moor" and reports to Holmes from what he calls:

> this most God-forsaken corner of the world. The longer one stays here the more does the spirit of the moor sink into one's soul, its vastness, and also its grim charm. When once you are out upon its bosom you have left all traces of modern England behind you.

This conclusion is hardly surprising because once you have entered the mire mother, or the mother of all mires, you have entered a pre-modern, even pre-Christian, landscape, or more precisely wetlandscape, which is the precursor of the modern landscape and the 'foundation' for it. Even though one may avoid sinking into it physically, it sinks into one metaphysically and psychologically. It is not a settled and stable object, but its own free agent, a dynamic process which cannot be kept within strict bounds, whose primary horror is that it breaks boundaries.

In the conclusion to the story Holmes and Watson follow Stapleton into "the widespread bog" with its "green-scummed pits and foul quagmires" where

> rank weeds and lush, slimy water-plants sent an odour of decay and a heavy miasmatic odour onto our faces, while a false step plunged us more than once thigh-deep into the dark, quivering mire [...] Its tenacious grip plucked at our heels as we walked, and when we sank into it, it was if some malignant hand was tugging us down into those obscene depths, so grim and purposeful was the clutch in which it held us.

Although Watson as a modern medical man does not seem to subscribe to a miasmatic theory of malaria, he supports a miasmatic theory of melancholia. He also anthropomorphises the mire by giving it hands to pull Holmes and him down into its clutching, (s)mothering embrace as it had pulled Stapleton "down in the foul slime of the huge morass."

2.7 The (V)alley of the Shadow of Death

Similarly, Fergus Hume in *The Mystery of a Hansom Cab* [hereafter *Hansom Cab*] first published in 1886, his even better-selling detective novel than *The Hound of the Baskervilles* and "the first popular Melbourne novel" (Arnold 1983, 7–8), set in part in the slums of nineteenth-century Melbourne, the detective, like the epic hero and the Christian pilgrim, enters the infernal regions displaced into the slums of the city figured as the biblical "valley of the shadow of death," an uncanny place. It evokes vividly the Dantesque "Infernal Regions" "off Little Bourke St" in the alley of the shadow of death. Also like the epic hero descending into the underworld to slay monsters and return home triumphant, "the detective led the way down a dark lane, which felt like a furnace owing to the heat of the night" (Hume 1999, 178–179).

The infernal distress of the slums of Bourke Street can be contrasted with what Marjorie Clark (in Arnold 1983, 69) in 1927 described as "the cool charm of its [Collins Street's] arcades." Worse for the detective entering the slums of Bourke Street, "it was like walking in the valley of the shadow of death [...] And, indeed, it was not unlike the description in Bunyan's famous allegory what with the semidarkness, the wild lights and shadows" (Hume 1999, 180). Hume combines Dante and Bunyan (and the biblical Psalmist, as Bunyan is drawing on Psalm 23) in an omnibus and ominous grab bag of tropes for the urban underworld.

As an aside here, these pages are excerpted and reproduced as a quintessential depiction of the "mean streets and back alleys" of late nineteenth-century Melbourne as depicted in crime fiction for the anthology *Literary Melbourne* edited by Stephen Grimwade (2009, 187–190). The reference to "the valley of the shadow of death" is later misquoted as "the shadow of the valley of death" (Grimwade 2009, 244). This misquotation gives a new nuance to the idea of death valley transposed here from the remote deserts of California to inner-city Melbourne. Both are hot and deadly places. As Michael Cannon (1976, 42 and 43) puts it, "the Angel of Death came early and stayed late in the Melbourne of 1892 and 1893" with "epidemics of influenza, typhoid and measles [...] which [...] killed thousands." It is only a hop, skip, and a jump from Bunyan's and the Psalmist's "valley of the shadow of death" to Hume's alley of the shadow of death and Grimwade's "shadow of the valley of death."

The detective's journey into the urban underworld is not only physical but also moral and allegorical. The detective novel is a secular allegory of literal and spiritual descent and ascent, of degradation and salvation, just like the sacred allegory of Bunyan's *Pilgrims Progress.* Flinders Street in 1845 was described as "that slough of despond" (cited by Annear 2014, 40). The swamp outside the city in the country, or inside the city in the muddy streets or slums, is a secularised satanic space. Perhaps it is fitting that the *flâneur* is a creature of the arcades, which Johann Geist (1983, vii) describes in his monumental history of the arcade as "a secularised sacred space." The swamp is a secularised satanic space for the detective that is a sacred space and place for indigenous, traditional peoples. The secularised satanic space of the swamp could be the native or natural swamp in the country, or the feral or cultural 'swamp' of the slums in the city in which the latter is used as a figure for the latter.

In *Pilgrim's Progress* Bunyan (2008, 65) takes the fourth verse of Psalm 23 referring to the valley of the shadow of death for his text and sermonises on how "Christian must needs go through it, because the way to the Celestial City lay through the midst of it." Typically the detective in the modern city must go through the (v)alley of the shadow of death of the underworld of the slums in order to protect the upper city of the upper ten thousand living in the upper world of the crystalline celestial city. Or the hero of the modern epic of the novel, such as Jean Valjean in Victor

Hugo's *Les Miserables*, must descend into the underworld of the sewers of Paris (see Giblett 2016, chapter 3). The modern hero of the tourist can even follow in Valjean's footsteps and go on a tour of the sewers of Paris in the marshy underworld of 'Lutetia,' the Latin-cum-Celtic name for the 'filthy marsh' in which the city was founded. Although Melbourne as the 'Paris of the South' does not have such an official tour of its sewers, "a small group has for several decades mounted regular unlawful explorations of Melbourne's stormwater drains and tunnels" in what Kristin Otto (2005, 175) calls "a Yarra [River] underworld." Sophie Cunningham (2011, 147) also comments that "while Melbourne does not boast Paris's hundreds of kilometres of underground tunnels, it does have Anzac, a cavernous drain under [the suburb of] South Yarra where many parties have been held over the years."[6]

Later in *Pilgrim's Progress* Christian is informed that "the valley [of the shadow of death] itself" is:

> as dark as pitch: we also saw there the hobgoblins, satyrs, and dragons of the pit; we heard also in that valley a continual howling and yelling, as of a people under unutterable misery, who there sat bound in affliction and irons; and over that hung the discouraging clouds of confusion; Death also does always spread his wings over it. In a word, it is every whit dreadful, being utterly without order. (Bunyan 2008, 67 and 69)

As the underworld is a monstrous place itself inhabited by monsters, such as dragons, so is 'the monster city.'

In similar vein, Little Bourke Street in *Hansom Cab* has its "weird and grotesquely horrible" inhabitants (Hume 1999, 220). In a word, they are monstrous. Little Bourke Street contrasts with Bourke Street proper, not only in the type of inhabitants, but also in the mode of illumination. Bourke Street is described "as the brilliantly lit street" produced by "electric lights," which highlight the members of the crowd caught in "the full glare of the electric light." By contrast, Little Bourke Street is lit by "sparsely scattered gas lamps" with their "dim light," whereas the lanes have no lights or lamps so they are dark, or "not quite dark, for the atmosphere had that luminous kind of haze so observable in Australian twilights, and the weird light was just sufficient to make the darkness visible."

[6] For further discussion of the cultural history and natural environment of Melbourne. "the Paris of the South," with its arcades and sluggish river like the Seine, see Giblett (2020).

(Hume 1999, 178–180). Typically the fictive detective of the modern city enlightens the benighted, brings light to darkness and illuminates the crepuscular gloom of crime and grime.

In her biographical and critical study of *Hansom Cab*, Lucy Sussex (2015, 7) argues that it had “an important role in establishing detective fiction as a publishing category” as Hume is “one of the most influential crime writers of all time” who wrote “the biggest-selling crime novel of the nineteenth century, and one of the most important Australian books ever.” Sussex traces the etymology of the word “detective” and how it literally means “de-roofing” as “a detective raises the roof, figuratively.” The detective in the detective story, and the detective storyteller, raises the roof of dwelling spaces, looks inside and reveals what is inside to the reader. The detective and detective story reader are positioned as snooping voyeurs.

Sussex (2015, 9–10) also traces how this uncovering and de-roofing had a demonic function and cites Dickens in *Dombey and Son* published in 1848 who pleaded for “a good spirit who would take the house-tops off.” David Grann (2017, 57) similarly traces how “the term ‘to detect”’ derived from the Latin verb to ‘unroof,’ and because the devil, according to legend, allowed his henchmen to peer voyeuristically into houses by removing their roofs, detectives were known as ‘the devil's disciples.’” In the Sherlock Holmes's story ‘A Case of Identity,’ Conan Doyle has Holmes describe the detective's work of unroofing in similar terms (albeit without the etymology and the theological overtones). The detective in the detective story, and the detective storyteller too, perform these socially useful, but morally ambiguous, roles of the good, or demonic, spirit. Both were *flâneurs* who entered “the city's central hell [of] the slums,” as Sussex (2015, 60) puts it in relation to Hume and *Hansom Cab*. As with the circles of Dante's hell, or inferno, the city has a centre, and its centre is the lower depth of its slums and sewers, the grotesque lower urban strata. The seemingly most secularised literary genre of the detective story about the modern city has loud theological undertones and those have strong philosophical underpinnings in the Greek philosophy of the elements and the humors.

The slough of despond still persists too. The figure is used with the same pejorative overtones as Bunyan (though not of course as an allegorical emblem for sin) in Donna Haraway's proclamation that “like Christian in *Pilgrim's Progress* [...] I am committed to skirting the slough of despond and the parasite-infested swamps of nowhere to reach more salubrious environs" (Haraway 1992, 295[7]). One wonders salubrious for whom? In whose terms? And if the slough of despond is not an allegorical emblem for sin, what is 'sin' in Haraway's theologised, or at least moralised, wet-landscape? What or where is 'nowhere'? Is it "the amniotic effluvia of terminal industrialism" (depicted on the cover of *Living with Nature in the Anthropobscene*) that Haraway refers to later (but without making any explicit connection)? If so, then the figure would have a critical edge and contemporary pertinence; if not, then Haraway's use of the figure would seem to be gratuitously uncritical about the use of tropes and blind to the politics of place (place-blind, placist and even misaquaterrist, hatred of wetlands).

Haraway has not heeded her own warning that she issued later in the same chapter when she argued that

> nature is not a physical place to which one can go, nor a treasure to fence or bank, nor an essence to be saved or violated. Nature is not hidden and so does not need to be unveiled. Nature is not a text to be read in the codes of mathematics and biomedicine. It is not the 'other' who suffers origin, replenishment and service. Neither mother, nurse, nor slave, nature is not matrix, resource, or tool for the reproduction of man. Nature is, however, a *topos*, a place, in the sense of a rhetorician's place or topic for consideration of common themes; nature is, strictly, a commonplace [...] nature is the place to rebuild public culture. Nature is also a *trópos*, a trope. It is figure, construction, artefact, movement, displacement. Nature cannot pre–exist its construction[...] Nature is a topic of public discourse on which much turns, even the earth [...] nature for us is *made*, as both fiction and fact [...] nature is made, but not entirely by humans: it is a co–construction among humans and non-humans [and the more-than-human]. (Haraway 1992, 296–297)

Nature is tropical figure and topical place. Yet the topes of nature are not troped equitably. Different *topoi* are troped in different ways with some being valorised at the expense or the detriment of others. There is a politics of the tropes of topes, a politics of *trópoi* of *topois*, such as with slimy swamps and marshes, and with slimy slums as swamps (as we will see in Chap. 9).

[7] Haraway (1992, 329, n.1) excludes the allegory of sin as slough of despond in quoting Bunyan's *Pilgrim's Progress*.

References

Annear, Robyn. 2014. *Bearbrass: Imagining Early Melbourne*. 2nd ed. Melbourne: Black Inc.

Aristotle. 1927. Problemata. In *The Works, Volume VII*, ed. W. D. Ross. Translated by E. S. Forster. Oxford: Clarendon Press.

Aristotle. 1930. De Generatione et Corruptione. In *The Works, Volume II*, ed. W. D. Ross. Translated by H. H. Joachim. Oxford: Clarendon Press.

Arnold, John. 1983. *The Imagined City: Melbourne in the Mind of Its Writers*. Sydney: George Allen and Unwin.

Bachelard, Gaston. 1983. *Water and Dreams: An Essay on the Imagination of Matter*. Translated by Edith R. Farrell. Dallas, TX: The Pegasus Foundation/The Dallas Institute of Humanities and Culture.

Beowulf and the Finnesburg Fragment. 1950. Translated by John R. Clark Hall, new edition. London: George Allen and Unwin.

Bunyan, John. 2008. *The Pilgrim's Progress*. Edited by Roger Pooley. London: Penguin. First published 1678.

Burton, Robert. 1932. *The Anatomy of Melancholy*. Edited by Holbrook Jackson. London: Dent. First published 1621.

Cannon, Michael. 1976. *The Land Boomers*. Illustrated ed. Melbourne: Thomas Nelson.

Creed, Barbara. 1986. Horror and the Monstrous-Feminine: An Imaginary Abjection. *Screen* 27 (1): 44–70.

Cunningham, Sophie. 2011. *Melbourne*. Sydney: New South.

Davison, Liam. 1993. *Soundings*. St Lucia: University of Queensland Press.

Douglas, Mary. 1966. *Purity and Danger: An Analysis of the Concepts of and Taboo*. London: Routledge.

Eagleton, Terry. 1986. *William Shakespeare*. Oxford: Basil Blackwell.

Forester, C. S. 1956. *The African Queen*. Harmondsworth: Penguin. First published 1935.

Fowler, Alistair, ed. 1994. *The Country House Poem: A Cabinet of Seventeenth-Century Estate Poems and Related Items*. Edinburgh: Edinburgh University Press.

Fritzell, Peter. 1978. American Wetlands as Cultural Symbol: Places of Wetlands in American Culture. In *Wetland Functions and Values: The State of Our Understanding*, ed. P. Greeson, J. Clark, and J. Clark, 523–534. Minneapolis, MN: American Water Resources Association.

Geist, Johann Friedrich. 1983. *Arcades: The History of a Building Type*. Cambridge, MA: Massachusetts Institute of Technology.

Giblett, Rod. 1996. *Postmodern Wetlands: Culture, History, Ecology*. Edinburgh: Edinburgh University Press.

Giblett, Rod. 2011. *People and Places of Nature and Culture*. Bristol: Intellect Books.

Giblett, Rod. 2016. *Cities and Wetlands: Return of the Repressed in Nature and Culture*. London: Bloomsbury.

Giblett, Rod. 2020. *Modern Melbourne: City and Site of Nature and Culture*. Bristol: Intellect Books.

Giblett, Rod. 2024. *Wetland Cultures: Ancient, Traditional, Contemporary*. New York: Palgrave Macmillan.

Grann, David. 2017. *Killers of the Flower Moon: Oil, Money, Murder and the Birth of the FBI*. London: Simon and Schuster.

Graves, Robert. 1955. *The Greek Myths*. Vol. II. Harmondsworth: Penguin.

Greenblatt, Stephen. 1982. Filthy Rites. *Daedalus* 111 (3): 1–16.

Grimwade, Stephen, ed. 2009. *Literary Melbourne: A Celebration of Writing and Ideas*. Prahran, Victoria: Hardie Grant.

Haraway, Donna. 1992. The Promises of Monsters: A Regenerative Politics for Inappropriate/d Others. In *Cultural Studies*, ed. Lawrence Grossberg, Cary Nelson, and Paula A. Treichler, 295–337. New York: Routledge.

Homer. 1965. *Odyssey*. Translated by Richmond Lattimore. New York: Harper and Row.

Hume, Fergus. 1999. *The Mystery of a Hansom Cab*. Melbourne: Text Publishing. First published 1886.

Huxley, Thomas H. 1989. Struggle for Existence. In *Mutual Aid: A Factor of Evolution*, ed. Peter Kropotkin, 329–341. Montreal: Black Rose. First published 1888.

Kant, Immanuel. 1952. *The Critique of Judgement*. Translated by James Creed Meredith. Oxford: Clarendon Press. First published in German in 1790.

Kant, Immanuel. 1960. *Observations on the Feeling of the Beautiful and Sublime*. Translated by John T. Goldthwait. Berkeley, CA: University of California Press. First published in German in 1764.

Kingsford, Richard. 1991. *Australian Waterbirds: A Field Guide*. Kenthurst, N.S.W.: Kangaroo Press.

Kirk, G. S., J. E. Raven, and M. Schofield, eds. 1983. *The Presocratic Philosophers: A Critical History with a Selection of Texts*. 2nd ed. Cambridge: Cambridge University Press.

Lloyd, G. E. R., ed. 1978. *Hippocratic Writings*. Translated by J. Chadwick, W. N. Mann, I. M. Lonie and E. T. Withington. Harmondsworth: Penguin.

Lyotard, Jean-François. 1991. *The Inhuman: Reflections on Time*. Translated by Geoffrey Bennington and Rachel Bowlby. Cambridge, UK: Polity.

Mattock, J. N., and M. C. Lyons, eds. and trans. 1969. *Hippocrates: On Endemic Diseases, Airs, Waters and Places*. Arabic Technical and Scientific Texts, 5. Cambridge: W. Heffer.

Merchant, Carolyn. 1980. *The Death of Nature: Women, Ecology and the Scientific Revolution*. San Francisco, CA: Harper & Row.

Mitchell, John G. 1992. Our Disappearing Wetlands. *National Geographic* 182 (4): 3–45.

Otto, Kristin. 2005. *Yarra: A Diverting History of Melbourne's Murky River*. Melbourne: Text Publishing.

Plato. 1954. Phaedo. In *The Last Days of Socrates*, 93–185. Translated by Hugh Tredennick. Harmondsworth: Penguin.

Ross, Sir David. 1995. *Aristotle*. 6th ed. London: Routledge.

Sir Gawain and the Green Knight [...]. 1975. Translated by J. R. R. Tolkien. London: George Allen & Unwin.

Sloterdijk, Peter. 1987. *Critique of Cynical Reason*. Translated by Michael Eldred. Minneapolis, MN: University of Minnesota Press. First published in German in 1983.

Sloterdijk, Peter. 2011. *Spheres, Volume 1: Bubbles: Microspherology*. Translated by Wieland Hoban. Los Angeles, CA: Semiotext(e). First published in German in 1998.

Sloterdijk, Peter. 2012. *The Art of Philosophy: Wisdom as Practice*. Translated by Karen Margolis. New York: Columbia University Press. First published in German in 2010.

Sloterdijk, Peter. 2013. *You Must Change your Life: On Anthropotechnics*. Translated by Wieland Hoban. Cambridge, UK: Polity. First published in German in 2009.

Sloterdijk, Peter. 2014. *Spheres, Volume 2: Globes: Macrospherology*. Translated by Wieland Hoban. Los Angeles, CA: Semiotext(e). First published in German in 1999.

Sloterdijk, Peter. 2016. *Spheres, Volume 3: Foams: Plural Spherology*. Translated by Wieland Hoban. Los Angeles, CA: Semiotext(e). First published in German in 2004.

Sloterdijk, Peter. 2020. *Infinite Mobilization: Towards a Critique of Political Kinetics*. Translated by S. Berjan.Cambridge, UK: Polity.

Sloterdijk, Peter. 2024. *Out of the World*. Translated by Corey A. Dansereau and Gill Zimmermann. Stanford, CA: Stanford University Press. First published in German in 1993.

Sloterdijk, Peter. 2025. *The Terrible Children of Modernity: An Antigenealogical Experiment*. Translated by Oliver Berghof. New York: Columbia University Press. First published in German in 2014.

Sussex, Lucy. 2015. *Blockbuster! Fergus Hume and the Mystery of a Hansom Cab*. Melbourne: Text Publishing.

Tillyard, E. M. W. 1963. *The Elizabethan World Picture*. Harmondsworth: Penguin. First published 1943.

CHAPTER 3

The Sublime and Slime

Abstract The sublime in chemical terms transforms the solid, or earth, into the gaseous, or air. The sublime for Zoë Sofoulis is always shadowed by slime, which she encapsulates in the parenthetical portmanteau s(ub)lime. Slime is the secret of the sublime. This chapter traces: the sublime in work of Longinus and slime in Dante's *Inferno*; through the sublime in the work of Kant and Burke, and slime in Sartre's *Being and Nothingness*; to the s(ub)lime in late twentieth-century critical theory, including Lyotard's, Berman's, and Eagleton's accounts of aesthetics and modernity. The sublime is central to western aesthetics, especially in the modern period. Around the name of the sublime, Lyotard remarks, "modernity triumphed," not least over wetlands in the short term by draining, filling, impounding, or embanking them, ironically using machines powered by fossil fuels from relict wetlands. The fossilised remains of swamps are sublim(at)ed into 'gas' to power machines and vehicles with steam, internal combustion, and jet engines in what Mumford called "carboniferous capitalism" that gave rise to the age of global boiling and the Anthropobscene. This chapter traces this irony, as well as critiquing the gender politics and ecoculture of the masculinist sublime and the misogynist feminisation of slime.

The concept of the sublime is central to western aesthetics, especially in the modern period. Indeed, modern aesthetics for Jean-François Lyotard

R. Giblett, *Living with Nature in the Anthropobscene*,
https://doi.org/10.1007/978-3-032-12822-5_3

(1984, 81) is "an aesthetics of the sublime." Elsewhere Lyotard (1989, 200) argued that "the sublime is perhaps the only mode of artistic sensibility to characterise the modern." Yet the concept is not of merely aesthetic interest and has a wider cultural pertinence to modernity in general and to the positioning of wetlands in particular. Around the name of the sublime, Lyotard (1989, 199) goes on to remark, "modernity triumphed," not least over wetlands in the short term, by draining, filling, or impounding them, ironically using machines powered by fossil fuels from relic wetlands. For Sloterdijk (2012, 166) "theory of modernity is always already, and only ever, possible as a reflection on the sublime in the manmade." The fossilised remains of swamps are sublim(at)ed to power machines and vehicles with steam, internal combustion, and jet engines in what Lewis Mumford (1934, 156–158) called "carboniferous capitalism" that gave rise to the age of global boiling and the Anthropobscene.

Given the centrality of the sublime to the project of modernity, it is hardly surprising to recall that the concept was important for both those colossi, and critics, of the modern, Marx and Freud. In *The Communist Manifesto* Marx and Engels used the metaphor of sublimation drawn from chemistry to argue that in bourgeois capitalism "all that is solid melts into air" (see Fig. 3.1):

Industrial capitalists and their technologies manipulated these transformative processes to greater or lesser good or harm. They transformed the elements into these combinations with sometimes devastating consequences, such as the firestorms of firebombing in World War II (as we will see in Chap. 8) and of feral bushfires in the age of global boiling and climate catastrophes in the Anthropobscene. Industrial capitalists and their technologies harnessed the power of steam to manipulate and manufacture other materials, such as metals from ores. They mined the solidified slime of carboniferous swamps when the world was wetland in the form of coal to power steam engines and generate electricity with both producing emissions of greenhouse gases to inaugurate global heating. Later they sublimated the oily slime of carboniferous swamps into the gas of diesel and petrol to power internal combustion engines with their emissions of greenhouse gases contributing to and increasing global heating. Then they fracked and desublimated natural gas into liquefied petroleum gas. These transformations lie at the heart of modernity for a couple of centuries. They are the basis for mainstream western culture for a couple of millennia with their foundation in the philosophy of the four elements.

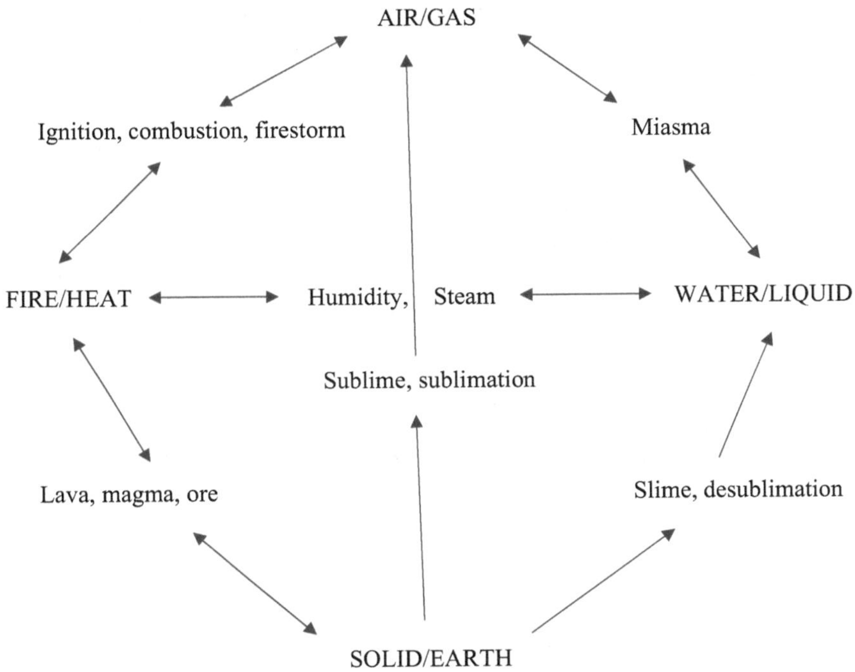

Fig. 3.1 The four elements and their transformations

"All that is solid melts into air" was taken up by Marshall Berman (1983) to encapsulate the etherealising project of modernity in which the solidities of pre-capitalist social formations are sublimed into thin air. The obverse of sublimation, or desublimation, or even detumescence, was used by what Klaus Theweleit (1987, 238) calls fascist soldier males to figure their fear of revolution in which "all that's solid becomes hot and fluid," becomes, in other words, a tropical swamp, an object, or more precisely to use Julia Kristeva's term, abject, of horror (Kristeva 1982). In opposition to fascism, and to sublimation and desublimation (insofar as it is tied to its obverse), what Elizabeth Grosz (1995, 203) calls "a mode of transubstantiation, a conversion from solid to liquid," has been used by her as a feminist figure for orgasm.

Freud (1985a, 39, 41 and 45) also used the metaphor drawn from chemistry to describe the process of sublimation by which sexual desire is displaced or deflected into ostensibly non-sexual realms, particularly the aesthetic and the intellectual.[1] The sublime and sublimation are closely linked as Kant (1960, 57) attests in an early work that the sublimation of "subduing one's passions through principles is sublime." The sublimation of sexuality into the aesthetic and intellectual can be seen as part of the condition of modernity, especially in the light of Georg Simmel's and Oswald Spengler's argument that the intellect developed in conjunction with the rise of the modern metropolis (Simmel 1950). Indeed, for Spengler (1932, 96; see also 92), "the city is intellect."

Yet, despite, or perhaps because of, the triumphs of modernity sublimation is always, as it were, haunted by its shadow, its 'other.' The sublime, Zoë Sofoulis suggests, is always shadowed by what Freud, following Schelling, calls the uncanny (discussed in Chaps. 4 and 5). Rather than the reverse of the sublime, for Sofoulis (1988, 12) "the uncanny is the obverse of the sublime, its other side: that from which it springs and that into which its turns," and even that into which it *re*turns (see Fig. 3.1). Sofoulis (1988, 15, n.86) goes on to suggest that the uncanny is associated with slime, and that "slime is the secret of the sublime," which she encapsulates in the parenthetical portmanteau "s(ub)lime." The home, or perhaps more precisely the 'unhome,' of the slimy, and the uncanny, is the wetland summed up in the reference to "the slime of the swamplands" (Tremayne 1985, 150).[2]

3.1 In the Slime with Dante and on the Sublime with Longinus

Although the slimy achieved a certain sort of modern philosophical cachet by virtue of receiving a ten-page treatment in Sartre's *Being and Nothingness* (to which I return below), it is not a category or experience which has received much attention in the western tradition despite its importance in *The Divine Comedy, Paradise Lost, Teenage Mutant Ninja Turtles*, and *Ghostbusters I* and *II*.[3] In the seventh Canto and the fifth

[1] See also Jean Laplanche and Jean-Baptiste Pontalis (1973).

[2] The association between slime and swamps (and other wetlands)—slimy places *par excellence*—is neglected or overlooked by Simon Estok (2024 in his ostensibly environmental humanities study of slime.

[3] Exceptions include Daly (1979, 332, 390n, and 402), Bauman (1973, 137–142), Collins and Pierce (1976), Harvey (1991), and Estok (2024). I am grateful to various friends, colleagues, and reviewers for their helpful suggestions of these and other references for me to follow up.

Circle of Dante's *Inferno* (and not the tenth level, as the Mayor of New York surmises incorrectly in *Ghostbusters II*) the condemned souls of the sullen, by a sort of sibilant (consonantal), alliterative (patriarchal) justice (at least in John Ciardi's English translation!), are stuck in the slime of the Styx, transformed from river to marsh by Dante, complete with black water:

> By that foul water, black from its very source,
> we found a nightmare path among the rocks
> and followed the dark stream along its course.
> Beyond its rocky race and wild descent
> the river floods and forms a marsh called Styx,
> a dreary swampland, vaporous and malignant.
> And I, intent on all our passage touched,
> made out a swarm of spirits in that bog
> savage with anger, naked, slime-besmutched.
> They thumped at one another in that slime
> with hands and feet, and they butted, and they bit
> as if each would tear the other limb from limb.
> And my kind Sage: 'My son, behold the souls
> of those who lived in wrath. And do you see
> the broken surfaces of those water-holes
> On every hand, boiling as if in pain?
> There are souls beneath that water. Fixed in slime
> they speak their piece, end it and start again:
> 'Sullen were we in the air made sweet by the Sun;
> in the glory of his shining our hearts poured
> a bitter smoke. Sullen were we begun;
> Sullen we lie forever in this ditch.'
> This litany they gargle in their throats
> as if they sang, but lacked words and pitch.'
> Then circling on along that filthy wallow,
> we picked our way between the bank and fen,
> keeping our eyes on those foul souls that swallow
> The slime of Hell. (Dante 1982, 55–56)

The slime of Hell is the product of the marsh or swamp of Hell, and the obverse of the sublime of 'Longinus.'

If slime is the secret or obverse of the sublime as Sofoulis contends, then the kind of qualities which Dante associates with the sullen stuck in the slime would have their obverse in the sublime. The earliest extant writing on the subject of the sublime in the western tradition dating from the first century of the common era is the treatise 'On the Sublime' attributed to Longinus (1965). In it 'Longinus' enumerates a number of qualities of the sublime which can quite readily be seen to have their obverse in the qualities Dante associates with slime. Let me list them: whereas the sublime for 'Longinus' exerts "an irresistible force and mastery" (100) (and thus can be considered masculine), slime for Dante exudes malignant vapours (and is considered feminine by Sartre); whereas the sublime for 'Longinus' "scatters everything before it like a thunderbolt" (100), slime for Dante engulfs everything in it like a nightmare; and whereas the sublime reveals in a flash "the full power of a speaker" (100), slime stops up in a mouthful the power of speech.

Moreover, whereas the sublime "uplifts our souls" (107), slime is the damnation of the sullen; whereas the sublime fills one with "a proud exaltation" (107), the souls of the sullen expel a base exhalation in slime; whereas the sublime is associated with a sense of "vaunting joy" (109), slime is associated with degrading wrath; whereas in the sublime "a noble emotion forces its way to the surface in a gust of frenzy" (109), in slime an ignoble emotion forces its way to the surface in an expulsion of gas; and whereas the sublime breathes "a kind of divine inspiration into the speaker's words" (109), the sullen in slime exhale a kind of devilish incantation in broken words.

In addition, whereas the sublime involves "the production of grand ideas, perpetually impregnating them with a noble inspiration" (109) (the sublime is a Bachelor Machine for a Bachelor Birth[4]), the sullen in slime produce base gutturals, perpetually aborting them with an ignoble afflatus; whereas the sublime involves catching fire from the inspiration of others (119), the sullen eat each other and the slime of hell; whereas the sublime impregnates one with heavenly power (119), the sullen in slime are inflated with a hellish cannibalism; whereas the sublime inspires one to speak oracles (119), the sullen in slime are inflated to emit noise; and finally, whereas the sublime carries one up close to the majestic mind of

[4] Klaus Theweleit (1987, 330, 414) has described Bachelor Births from Bachelor Machines as "attempts to create a new reality by circumventing the female body – to engender the world from the brain" and from the head as the site of the superego, both like Zeus giving birth to Pallas Athene from his head, arguably the first Bachelor Birth from a Bachelor Machine and the first 'brainchild.' For Bachelor Machines, see also Michel de Certeau (1984, 150–153, 1986) and Michel Carrouges (1954). For Sloterdijk (2017, 71) glossing Mircea Eliade "the birth of technology [comes] out of the spirit of masculine substitute pregnancy."

God (147), the sullen in slime are carried away close to the debased lusts of the Devil.

3.2 On the Sublime and Beautiful with Kant and Burke

The concept-metaphor of the sublime is later taken up by Burke and Kant who rework it, especially in relation to the sensations with which it is associated, though it is still inevitably haunted by its slimy 'other.' For Burke (1958, 57) the sublime is a delight which turns on pain. It also causes astonishment, "that state of the soul in which all its motions are suspended with some degree of horror," though for Kant (1952, 120) the astonishment of the sublime amounts almost to terror (see also Kant 1960, 48). By contrast, the uncanny for Freud is related to what is frightening and to what arouses "dread and horror" (Freud 1985b, 339). Whereas the sublime is attended with "some degree" of horror, the uncanny arouses exclusively dread and horror. The ruling principle of the sublime for Burke is terror (Burke 1958, 58), whereas the ruling principle of the uncanny for Freud is horror.

Terror and the sublime, pleasure and the beautiful, could be said to be the ruling aesthetic responses and cultural passions of modernity, whereas horror and the uncanny may be those of hypermodernity. The sublime implies an impossible object[5] (such as dead rock) terrifying in its magnitude, whereas the uncanny (and slimy) intimates an impossible abject (such as living water) horrifying in its monstrosity and the beautiful implies a possible object pleasing in its smallness (such as a gentleman's park or a feminine woman). The first is arguably associated with the exteriority of the Law of the Father, whereas the second is linked with the interior of the Body of the Mother and the third with the shape and surface of the earth

[5] Against the sublimes of Kant and what he calls the "post avant-gardes" Dick Hebdige (1987) valorises the vulgar, the popular and 'the body' (though whose body? one might ask). Yet repugnance for or valorisation of the vulgar, the popular and 'the body' can be regarded as a sterile antinomy which occludes how the sublime is supported precisely by sublimation of the vulgar, the popular and 'the body.' Part of the impossibility of the sublime object lies not only, as Hebdige suggests, in being in a sense beyond the fetishisation of commodities, but also in its production from raw materials by labour and its imbrication in the circuits of sublimation. The vulgar, the popular, and 'the body' do not have a fixed and necessary meaning or resistive force against capitalism (early or late), but are sublimated in the production and consumption of commodities.

of idealised land or of the body of the idealised woman in patriarchy. The distinction between the beautiful and the sublime has been gendered with the former being ascribed to the female/feminine and the latter to the male/masculine (Kant 1960, 77 and 93).

Yet the sublime, terror, horror and the uncanny are not just mere aesthetic responses, nor are they qualities of objects but modalities which set up and involve relations between subjects and objects, or abjects in the case of the uncanny. The sublime is a feeling of awe and fear in the face of symbolic objects, such as mountains, insofar as they are symbolic of the ultimate symbolic object, the Phallus and the Law of the Father. For Kant (1960, 47 and 1952, 121) the sight of a mountain evokes a feeling of the sublime. Similarly for Henry David Thoreau (1980, 44) "sublimity and grandeur [...] belong to mountain scenery." John Muir (1987, 110, 111, 115, 122, 140, 401) repeatedly referred to "mountain sublimities."

The quintessential pictorial evocation of mountainous subilimity is 'The Wanderer' by Caspar David Friedrich with its individualist nomad/monad astride the pinnacle surveying the cloud-swept, sea-like heights (reproduced on the cover of Eagleton 1990). Sloterdijk (2013b, 443) defines the sublime as "that which, by calling to mind the overwhelming, shows the observer the possibility of their engulfment by the oversized – which, however, is suspended until further notice." By contrast, the uncanny is a feeling of horror in response to imaginary abjects, such as smells, insofar as they are images of the ultimate imaginary abject, the Body of the Mother. Finally, horror is an emotion comprising fear and loathing in response to more immediate contact, especially via the sense of touch, with the Body of the Mother, particularly with its inside and with those regions of the earth associated with it such as the swamp.

The beautiful, though, is quite different. For Burke (1958, 113), the sublime dwells on great and terrible objects, whereas the beautiful dwells on small ones and pleasing. Similarly for Kant (1960, 48) "the sublime must always be great; the beautiful can also be small. The sublime must be simple; the beautiful can be adorned and ornamented." Whereas the sublime has, or is associated with, a broken and rugged surface, the beautiful for Burke has, or is associated with, a smooth and polished surface (Burke 1958, 114). Whereas the sublime concerns or represents vast objects, the beautiful, comparatively small objects. Whereas the sublime is dark and gloomy, the beautiful is not obscure (124). Whereas the sublime is founded on pain, the beautiful is founded on pleasure and acts by relaxing the solids of the whole system and so is the cause of all positive pleasure (50).

The beautiful is a kind of enemetic evacuation or spermatic ejaculation, whereas the sublime could be characterised as 'a good shit' of solid objects, or turds. Thus Gilles Deleuze and Felix Guattari (1977, 143) have remarked that "sublimation is profoundly linked to anality."

3.3 In the Slime with Sartre

During the course of what he calls his "psychopolitical investigation" into rage and time, Sloterdijk (2010, 176) suggests that Jean-Paul Sartre, "the author of *Being and Nothingness* and the *Critique of Dialectical Reason*," is "one of the greatest intellectuals of the [twentieth] century."[6] I suggest that Sartre's *Being and Nothingness* is probably the greatest proof text from the twentieth century for a critique of the masculinist gender politics of slime in a psycho- and femo-political ecology of the Anthropobscene.[7] Sloterdijk (2013a, 91) comments three years later and 30 years after Sartre's death in 1980 that he "already appears like a monumental figure in the history of modern literature and philosophy." Indeed, Sartre was a colossus on the side of the patriarchal paradigm of the phallic monumental that represses and oppresses the matrifocal paradigm of the monstrous maternal (Giblett 2011, Figure 1). Sloterdijk (2013a, 94) concludes that Sartre was "perhaps the most diligent, active philosophical author of the twentieth century." Sartre indeed was active and an activist. He did not practice Taoist non-action (*wu-wei*) and he was loath to go into the watery places loved by the Tao, or the way, and hated by some men, such as slime (as we will see in Chap. 4). When he did go into slime, he carried his cultural baggage with him — and he was horrified. He had not embraced feminism, nor had he engaged critically with the gendered construction of reality before he ventured into slime metaphorically speaking.

Slime is construed as feminine in Sartre's *Being and Nothingness*. In about ten markedly misogynist pages of his *magnum opus*, his big dick/gun/work, Sartre (1969, 611) undertakes a philosophical meditation on what he calls "the horror of the slimy." For Sartre, the slimy is feminine and an object of horror for what he calls "the European adult" (604), though I would suggest he is referring to the white male adult. The white

[6] Jean-Pierre Couture (2016, 8–30) argues that "Sloterdijk's 'psychopolitical inquiry emerges out of his doctoral thesis." It continued with his founding of psychopolitical spherology.

[7] For definition and discussion of femopolitical ecology, see Giblett (2026).

male adult is constituted as such by *not* being the black feminised water of the wetlands and by being horrified by the (s)mothering body of the swampy mother.

In Sartre's own philosophical terms, the slimy represents the revenge of the in-itself, or solid exemplified in "the immanent body," on the for-itself, or fluid idealised in "consciousness and reflexive self-consciousness."[8] Part of the horror of the slimy resides in the fact that it is neither solid, nor fluid, as part of the horror of the swampy lies in the fact that it is neither earth, nor water. They are neither one thing nor the other, but occupy an unsettling middle position between two things. They both fall into, or are more precisely, a mediating category between the two extremes of solid earth and liquid water (see Fig. 3.1). Yet for Sartre the in-betweenness of the slimy does not represent a settled state or a stasis. Rather, the slimy represents for Sartre the solidifying of the fluid, the damming of its flow, the damning of the flows. For him "a slimy substance is an aberrant fluid," a fluid which does not flow in the normal channels.

As slime is an aberrant fluid, it does not let itself be contained, and it does not allow one to make a mark on it, or more precisely, to make one's mark on it. For Sartre (1969, 607), the slimy is a becoming "on which one leaves no mark, and which could not leave a mark on us." The slimy, like, or more precisely *as*, the swamp, does not allow the writing of development to take (a) place on the earth unless and until it is filled or drained. It does not allow unaltered any fixtures which are permanent. For Sartre, the slimy is "a being which is eternity and infinite temporality" (607). It does not permit the stasis of western 'man-made' structures.

The slimy, however, is ambiguous because it represents for Sartre, at least partially, "a dawning triumph of the solid over the liquid" (607), as the swamp of the Bora in C. S. Forester's *The African Queen* "was a region in which water put up a good fight against the land which was slowly invading it" (Forester 1956, 136), a fight which nevertheless it is doomed to lose. On the one hand, the slimy and the swamp obstruct the solid from taking (a) place, and on the other, the slimy, and the swamp, are the invasion and triumph of the solid earth over the liquid water. It is this undecidability of the slimy that makes it such an object of horror. It is neither solid nor liquid, but it just does not sit still as some sort of fixed and static mediator. It is both the process of one becoming the other (the liquid

[8] See Grosz (1994, 86).

solidifying), and the prevention of the one taking the place of, or taking place on, the other (the solid being erected on it). For Sartre, the slimy is:

> a tendency of the indifferent in-itself which is represented by the pure solid, to fix the liquidity, to absorb the for-itself which ought to dissolve it. Slime is the agony of water. It presents itself as a phenomenon in process of becoming. (Sartre 1969, 607)

Hence the slimy is horrific because it is not being but becoming.

In its becoming, the slimy for Sartre represents what he calls "a constant hysteresis in the phenomenon of being transmuted into itself" (608). As the French feminists have reminded us, the word 'hysteria,' and its derivatives such as hysteresis, come from the Greek word 'hysteron' meaning 'womb.' Sartre's uses the word "hysteresis" pejoratively to refer to matter being transmuted into itself in the womb, rather than into something other than itself; matter enwombed in slime to produce more matter rather than matter sublimated into the sublime to produce non-matter. Sartre's use of the word "hysteresis" can also be read historically in J. J. Bachofen's terms as a patriarchal supersession of the matriarchal, of the swampy and slimy womb of life, or "the lustful chaos of primeval swamps" as Gilles Deleuze (1989, 52–53) puts it, by the agricultural.

The becoming of the slimy contrasts strongly with the being of the solid whose inertia for Sartre "symbolises for me my total power" (Sartre 1969, 608). The solid is prototypically the turd which is ideally not so soft and malleable to be slimy, nor so hard and dry to be intractable. The solid turd is also prototypically the first form of the gift from the child to the mother whereas the slimy placenta is the first gift from the mother to the foetus (Rouch cited by Luce Irigaray 1993, 43–44). Sloterdijk (2011, 376–377) traces the etymology of "placenta" from the Latin for flat cake or flat bread. The English slang for pregnancy of having 'a bun in the oven' is part of the same lineage. Sloterdijk (2011, 377) notes, however, that "according to early traditions of midwifery, [...] the dough baking in the maternal oven was not so much the child itself as that mysterious placental cake on which the child [...] fed *in utero*." His translator notes that "one German word for the placenta, and the one used in this sentence, is *Mutterkuchen*, 'mother cake'" (Sloterdijk 2011, 649 n12).

The placenta is a more primary gift than the turd — the placenta is the first gift. The qualities of objects have implications for the power which can be wielded over them. Whereas the solid symbolises for Sartre his total power, presumably, by implication, the slipperiness of the slimy, its aberrance, symbolises for him a loss or defeat of mastery. As I have already remarked, the sublime is the pain/pleasure of a 'good shit,' and the beautiful is a kind of enematic evacuation, or spermatic ejaculation, whereas the slimy is a slipping away, perhaps associated with diarrhoea, but also a sucking in.

Power over the turd resides in being able to expel or eject it whereas the horror of the slimy consists in that it can ingest. The slimy for Sartre, like the swamp in patriarchal western culture I would add, (and here I quote at some length):

> sticks to me, it draws me, it sucks at me. Its mode of being is neither the reassuring inertia of the solid nor the dynamism like that in water which is exhausted in fleeing from me. It is a soft, yielding action, a moist and feminine sucking, it lives obscurely under my fingers, and I sense it like a dizziness; it draws me to it as the bottom of a precipice might draw me. There is something like a tactile fascination in the slimy. I am no longer the master in arresting the process of appropriation. It continues. In one sense it is like the supreme docility of the possessed, the fidelity of a dog [and a woman ?] who gives himself [herself?] even when one does not want him [her?] any longer, and in another sense there is underneath this docility a surreptitious appropriation of the possessor by the possessed. (Sartre 1969, 609)

The slimy upsets the solidities of the positions of master and possessor, and indeed threatens to reverse the roles of possessor and possessed.

Sartre then goes on to describe this possession by the possessed of the possessor as "a poisonous possession" (609), whereas presumably by implication the possession by the possessor of his possessions, his solid objects, his turds, is not. Sartre then switches metaphor to talk about "the snare of the slimy: it is a fluidity which holds and which compromises me, [...] it clings to me like a leech [...] and leaves its traces upon me" (609). Although one cannot leave a mark on the slimy as William Byrd found with the Great Dismal Swamp,[9] it leaves its traces on one as Sartre says. The swamp not only "attacks the surfaces of the body," as David Miller (1989, 178) as puts it, but also marks them. Rather than being a surface for inscription, the slimy and swampy mark the surfaces of the body. They not only represent an impossible writing surface, but is also a, if not the,

[9] For William Byrd and the Great Dismal Swamp on the border between Virginia and North Carolina, see Giblett (1996).

primal writing of the trace. For Sartre (1969, 609), "the slime is like a liquid seen in a nightmare [a particular kind of masculinist nightmare, I might add] […] Slime is the revenge of the In-Itself. A sickly-sweet feminine revenge." The slimy leaves its traces across most of the senses — it tastes sickly-sweet, it looks liquid (or solid) but isn't really, it feels fascinating, but "to touch the slimy is to risk being dissolved in sliminess" (610).

Slime is not only problematic spatially but also temporally. Whereas for Sartre (1969, 610) following Bergson, the slimy is associated with eternity and infinite temporality, the duration of consciousness is like a river. As the slimy is construed as eternity and infinite temporality and marked as feminine, so for Sartre the river is duration of consciousness and marked as masculine. To the list of qualities which have been ascribed to the masculine and the feminine in patriarchal western culture we can add a couple more:

Activity	Passivity
Culture	Nature
Father	Mother
Head	Heart
Man/Masculine	Woman/Feminine
Reason	Emotion
Action	Speech
Doing	Be(com)ing
River	Swamp
Flow	Stagnation
Sublime	Slime
Duration of consciousness and infinite spatiality	Eternity and infinite temporality[10]

Sartre then goes on to associate the slimy with holes, holes in the sand and in the earth (612), though he does not mention wetlands. The function of a hole for Sartre is to be filled and this filling is construed in masculinist sexual terms. For him, "the hole is originally presented as a nothingness 'to be filled' with my own flesh" (613). Fucking is construed here in masculinist terms of filling holes, whether it be holes in human

[10] This list is based on, and add to, those of Hélène Cixous and Catherine Clément (1986, 63) and Cate Poynton (1985, 18).

bodies (though which holes in whose body Sartre does not specify), or holes in mother earth. For Sartre:

> the ideal of the hole is then an excavation which can be carefully moulded about my own flesh in such a manner that by squeezing myself into and fitting myself tightly inside it, I shall contribute to making a fullness of being exist in the world. (613)

Sartre's desire for a snug hole does not differentiate between anus and vagina, and so includes the anal eroticism of fucking the arseholes of mother earth. The slimy and the swampy in matriarchy had been the womb of life whereas in patriarchy they have become the arsehole of the earth where wastes are excreted or dumped. Patriarchy then fails to differentiate between the two regimes and regions and treats the former as it does the latter. Yet Sartre's desire for a snug hole is also a recognition of lack and a desire for union and completion, if not for procreation.

Sartre then triumphantly announces that "here at its origin we grasp one of the most fundamental tendencies of human reality [at least of a masculine construction of reality] — the tendency to fill" (613). If anything could be said to characterise the history of wetlands in western culture it would be this tendency to fill, or more precisely, as "tendency" is a euphemism if there ever were one, what could be called the will to fill (as we will see shortly).

Sartre goes on to say with boyish good humour that "a good part of our life is spent in plugging up holes [like the Dutch boy plugging up the dyke], in filling *empty* [my emphasis] places [like wetlands], in realising and symbolically establishing a plenitude" which produces a full, replete and smooth surface and which destroys by filling. As slime or wetland is to the unconscious, so dykes and drainage are to sublimation (as we will also see shortly). Sartre then goes on to blithely say that "it is only from this standpoint that we can pass on to sexuality" (613). As if sexuality were not already present in his desire for a snug hole and his desire to anal-fuck the earth. What he seems to be implying is that it is only from the standpoint of the will to fill that he can pass on to a discussion of *female* sexuality, so implicitly admitting that he has been concerned with *male* or masculine sexuality as normative.

Whereas Freud associates the uncanny with female genitalia (as we will see in Chap. 5), Sartre associates the slimy with them. For both of them uncanny and slimy female genitalia are disturbing and disrupting. For

Sartre, "the obscenity of the feminine sex is that of everything which "gapes open." It is *an appeal to* being as all holes are" (613; his emphases). In other words, it is not being but a lack of being. Yet is not equally Sartre's desire for union an appeal to being? For Sartre it seems, as for the fascist soldier male, "women's bodies are the holes, swamps, pits of muck that can engulf," as Barbara Ehrenreich puts it in her 'Foreword' to volume one of *Male Fantasies* (Theweleit 1987, xiii). But women's bodies could only be seen as such as the result of confusing anus and vagina/womb (the excretory and reproductive/sexual functions) and so denying sexual difference (with women because men have an anus too)[11] and simultaneously polarising gender in non-reproductive terms (across men and women). By desiring indiscriminately a snug hole, and so failing to differentiate between vagina/womb as source of life and anus as site of waste, Sartre denies sexual difference with women, especially women's capacity to procreate.

At the same time, by associating women's bodies with slime and holes, and by associating men's bodies with solid objects, Sartre polarises gender in non-reproductive terms across men and women. This is a typically fascist move as Jessica Benjamin and Anson Rabinbach point out in their 'Foreword' to volume two of Theweleit's *Male Fantasies* (1989, xix).:

> fascism is an extreme example of the political polarisation of gender with two basic types of bodies [which] exemplify the corporeal metaphysics at the heart of the fascist perception: the soft, fluid, and ultimately liquid female body which is [exemplified in the swamp and is] a quintessentially negative 'Other' lurking inside the male body [... and] the hard, organised, phallic body devoid of all internal viscera which finds its apotheosis in the machine,

whose essential characteristic is, as de Certeau (1986, 166) has remarked, 'male,' though of course not biologically male. The Existentialist Marxist Sartre and the fascist soldier male share a common gender politics. Sartre is a prime twentieth-century example of a philosopher-sublimator with his misogyny and masculinist gender politics of slime. The philosopher is not a revolutionary desublimator.

In response to the project of capitalist modernisation in which "all that is solid melts into air" and to the threat of communist revolution in which

[11] See Kristeva (1986, 147).

"all that is solid becomes hot and liquid," for the Fascist soldier male as Ehrenreich puts in her 'Foreword' (Theweleit 1987, xv):

> all that is rich and various must be smoothed over (to become like the blank facades of fascist architecture); all that is wet and luscious must be dammed up and contained; all that is 'exotic' (dark, Jewish) must be eliminated.

Or in the terms of the sublime, the beautiful and the slimy, all that is rich and various, wet and luscious, exotic and dark like the wetland must be given the smooth surface of the beautiful (like the canalised wet lands of John Constable's wet landscapes) and not be rendered into the rough surface of the sublime, nor be permitted to remain the sticky, sucking surface, and depths, of the slimy and the swampy.

The project of capitalist modernity takes the "base matter," or "slimy bads" as Zoë Sofoulis calls them, of the earth and transforms or sublimates it into 'commodities,' or what Sofoulis calls "shiny goods" of industrial capitalist technology, then dumps its waste products back into the slimy, swampy places of the earth in the Anthropobscene, and so the process comes full circle back, as it were, to where it began (see Giblett 1996, Figure 1 and the cover of *Living with Nature in the Anthropobscene*). But in so doing the earth is destroyed to become like the landscape of trench warfare which Walter Benjamin saw as the terrain of idealism and which can be seen as the final outcome of sublimation:

> it should be said as bitterly as possible: in the face of 'this landscape of total mobilization,' the German feeling for nature has had an undreamed-of upsurge. The pioneers of peace, who settle nature in so sensuous a manner, were evacuated from these landscapes, and as far as anyone could see over the edge of the trench, the surroundings had become the terrain of German idealism; every shell crater had become a problem; every wire entanglement an antinomy; every barb a definition; every explosion a thesis; by day, the sky was the cosmic interior of the steel helmet, and at night, the moral law above. Etching the landscape with flaming banners and trenches, technology wanted to recreate the heroic features of German Idealism. It went astray. What it considered heroic were the features of Hippocrates, the features of death. Deeply imbued with its own depravity, technology gave shape to the apocalyptic face of nature and reduced nature to silence – even though this technology had the power to give nature its voice. War, in the metaphysical abstraction in which the new nationalism believes, is nothing other than the attempt to redeem, mystically and without mediation, the

> secret of nature, understood idealistically through technology. This secret, however, can also be used and illuminated via a technology mediated by the human scheme of things. (Benjamin 1999, 318–319)

The process of sublimation can be seen here to have returned full circle, as it were, to the slime from which the whole process could be seen to have begun, so that it can begin again, though it has never stopped (see Fig. 3.1 and Giblett 1996, Figure 1). Benjamin brought philosophy, ecoculture, and psychopolitical ecology together here in an exemplary, creative, and critical manner. He also did so in other remarks about World War I (as we will see in Chap. 7).

Sloterdijk (1987, 114) aptly calls Benjamin "that great expert in polyvalent meaning who built secret bridges between Judaism and sociology, Marxism and messianism, art and critique." Sloterdijk does not mention that Benjamin also built not so secret bridges between philosophy and all of these '-isms,' aspects, and approaches. Twenty-five years later Sloterdijk (2013b, 136) does not include Benjamin among those who transformed philosophy "back to an engaging discipline."[12] Arguably Benjamin transformed philosophy forward into engaging with other disciplines, such as history and literature in transdisciplinarity (as Sloterdijk does too).[13] Rebecca Solnit (2000, 197, 198) acknowledges that Benjamin was a transformative philosopher, "the patron saint of cultural studies" and "one of the great scholars of cities and the art of walking them." She also calls him "the incomparable, uncategorizable, German-Jewish essayist and theorist" and "the incomparable writer-philosopher" who "in his work was a historian, a theorist, a lyrical writer, creating writing as uncategorizable as it is influential" (Solnit 2005, 99, 2007, 69, 70). Benjamin was transdisciplinary before the term was invented. He let philosophy out of its disciplinary prison to roam free.

Benjamin also built, crossed, and destroyed a not-so-secret bridge between sociology and cultural studies. "Sociology was invented to forget politics," as Paul Virilio and Lotringer (1983, 10) said. Cultural studies was invented to remember politics; Benjamin was at the forefront of this invention (before Raymond Williams and Stuart Hall). Ecocultural studies was invented to remember the interactions between human beings and the earthly household; Benjamin and Williams were at the forefront of this

[12] They are Nietzsche, Wittgenstein, Heidegger and Foucault.
[13] As Nietzsche and Foucault did for Sloterdijk.

invention. Ecofeminism was invented to remember the gender politics of nature; many feminists and other women were at the forefront of giving birth to it in a collective and collaborative labor (Giblett 2026).

Psychoanalytic ecology was invented to remember the psychodynamics (including psychopathology) of living beings living on the living earth, on a continuum between psychogeopathology and psycho-symbiosis (Giblett 1996, 142, 2019). The invention of psychopolitical ecology remembers the politics and psychodynamics (including psychopathology and psycho-symbiosis) of living beings living in the earthly household (*oikos*). Benjamin is at the forefront of this invention too, as is Sloterdijk with what he calls "a political poetics of space," including "psychodynamic space, [...] the space in which existence takes place" (Noordegraaf-Eelens and Schinkel 2011, 185; Sloterdijk 2014, 951). In the less abstract and more concrete terms of place rather than space, psychopolitical ecology advances the home habitats of bioregional places within the biosphere of living beings and the earthly household of the ecosphere.

Henry David Thoreau was at the forefront of this invention with his insistence in his journal that "home is here" (Thoreau 1962, *V*, 254). He elaborated in his journal several years later:

> Here I am at home. In the bare and bleached crust of the earth I recognize my friend [...T]he constant endeavour should be to get nearer and nearer *here* [...] A man dwells in his native valley like a corolla in its calyx, like an acorn in its cup. *Here*, of course, is all that you love, all that you expect, all that you are. Here is your bride elect, as close to you as she can be got. Here is all the best and all the worst you can imagine. (Thoreau 1962, *XI*, 275; his emphases)

Home is here; it is not there, some place elsewhere. Thoreau wrote earlier in his journal that "the man who is often thinking that it is better to be somewhere else than where he is excommunicates himself" (Thoreau 1962, *X*, 190–191). "He" is no longer in communication with the sacred place of home, but has banished and excommunicated "himself" into exile.

3.4 On the Sublime with Eagleton and in the Swamp with Freud

Psychopolitical ecology includes the ecology of gendered bodies and the gendered construction of reality. Critical attention to the gender politics of the sublime has been given by Terry Eagleton, though without discussing the gender and environmental politics of the metaphors he employs to do so whose burden is, as always, and as has been remarked by Kay Schaffer (1988, xii), carried by women — metaphor is women's burden — and by mother earth. Luce Irigaray (1985, 140, 144, 229, and 263) likewise refers to the "'masculine' games of tropes and tropisms" in which women are "rolled up in metaphors, buried beneath carefully stylized figures," though "no metaphor completes her" despite the fact that "the womb has been played with, made metaphor and mockery of by men" (see also Derrida 1982). Eagleton shows how the sublime in the western tradition is associated with the masculine virtue of "virile strenuousness." Unlike the femininity of beauty, the sublime is, Eagleton (1990, 54) argues, "on the side of enterprise, rivalry and individuation". Yet the sublime and the beautiful do not co-exist side-by-side in blissful harmony. Rather, the sublime quite literally colonises the beautiful which, in turn, attempts to colonise the slimy. Eagleton (1990, 54) argues that the sublime, "the lawless masculine force [...,] violates yet perpetually renews the feminine enclosure of beauty" which, in turns, encloses the uncanny, the slimy and the swampy.

Yet enclosure is no mere metaphor. In the British enclosure movement of the eighteenth century, as Raymond Williams (1985, 96) puts it, "by nearly four thousand Acts, more than six million acres of land were appropriated" (though much of this land was specifically *wet*land, a crucial distinction). Enclosure, argues E. P. Thompson (1968, 237), was "a plain enough case of class robbery." The enclosure movement in wetlands was invariably associated with drainage in order to render the land profitable for monoculture with, as Christopher Hill (1965, 260) maintains, "fen drainage alone extending the country's arable area by 10%." In fact, the new outburst of drainage projects of the 1760s was contemporaneous with and supportive of the beginning of the agricultural revolution (Thirsk

1957, 205) over the local, native paludiculture with the cultivation of wetlands.[14]

Enclosure is to the beautiful as drainage is to sublimation. Both enclosure and drainage were exercised against the slimy. For Freud (1964, 80) the culture-work of his famous and oft-quoted pronouncement of "where id was, there ego shall be" is, in the terms of his less quoted analogy, "not unlike the draining of the Zuider Zee." The draining of the Zuider Zee was probably the single-largest drainage project ever undertaken, amounting to over half a million acres (Van Veen 1949, 52). The devastating effects of this project on the local people, their bioregion and livelihood are graphically depicted in the film *Doodwaater*, literally 'dead water.' According to Sloterdijk (2024, 19) tens of thousands of windmill-driven water-pumps "enlivened the landscapes of northern European countries from the sixteenth century onwards." They certainly did so for agriculture and landscape painting within the aesthetic conventions of the picturesque, but they also deadened the wetlands of these countries for paludiculture, as well as anesthetised and aestheticised wetlandscape painting, with some possible exceptions such as Peter de Wint, John Sell Cotman and recently Nick Tearle (see Giblett 2024, chapter 6).

Sloterdijk (1987, 366; 380, n9) quotes Freud's pronouncement and critiques the analogy in an endnote as:

> strictly speaking erroneous; it is still too colored by engineering, conceived too much according to the model of polemical medicine (sickness as enemy). The id-sea never dries up – but is it not already a lot when one has learned to sail on it?

According to the model of holistic mind-body medicine (sickness and health as life), the id is the fertile and slimy swamp of the uncanny (as we will see in the following chapters). Freud's pronouncement thus needs to be rewritten: where ego is, there id was and shall still be, not unlike the draining of the English Fens. They were never dried up and the id-swamp never dries up. Is not the id-swamp already a lot when one has learned to wade into it?

Dyking the Zuider Zee produced dead water; attempts at draining the id produce the dead water of the dammed, and damned, dry land of the

[14] For further discussion of the enclosure and drainage of the English Fens, see Giblett (2025, chapter 6). For further discussion of paludiculture, see Giblett (2025).

bourgeois ego, unlike the living black water of the wild wetlands of indigenous cultures. Opening the floodgates of the id created what Sloterdijk (2025) calls "the super-id" of "the terrible children of modernity" (Napoleon, Hitler, Stalin, etc.). The super-id excreates the feral dead black waters of wastelands in the Anthropobscene (see the cover of *Living with the Anthropobscene*). What Byung-Chul Han (2017, 63–65) calls the psychopolitics of the new technologies of power open the floodgates of the id in what he also calls "the digital unconscious." The digital unconscious is the breeding ground for the super-id of the terrible children of the hypermodernity with their technologies of power.[15] Han is drawing on Benjamin's concept of "the optical unconscious" in early photography in which "another nature speaks to the camera" (Benjamin 1979, 243, 1999, 510, 2003, 266). With the digital unconscious computational capitalist culture speaks to the camera, the screen, the eye, and the I; nature is othered, the I is the same.

Freud would have preferred that the id did dry up completely as his erroneous pronouncement indicates. The id needs to be drained for Freud (1964, 73) because it is "a chaos, a cauldron full of seething excitations" which run into it from the instincts. If culture-work is ultimately draining instinct, then the threat of instinct for Freud (1973, 353) is that it "would break down every dam and wash away the laboriously erected work of civilisation" and we would flounder about, as Lyotard (1993, 58) puts it, in the swamp of uncertainty that fashions the instincts themselves. Freud was both an engineer of the modern mind and an avatar of the ancient uncanny, his contradiction, and fascination, like the uncanny (as we will see in a later chapter).

For Kant too, the sublime is associated with a bulwark against the conscious being swamped by the unconscious. The Kantian sublime is, Eagleton argues:

> in effect a kind of unconscious process of infinite desire which like the Freudian unconscious continually risks *swamping* and overloading the pitiable ego with an excess of affects. The subject of the sublime is accordingly decentred, plunged into loss and pain, undergoes a crisis and fading of identity; yet without this unwelcome violence we would never be stirred out of ourselves, never prodded into enterprise and achievement. We would lapse

[15] To name names at this point would be defamatory under US law and against the policy of the publisher. I leave it to the imagination of the reader to fill in the blanks of the global purveyors of the super-id and their and your own local acolytes.

> back instead into the placid feminine enclosure of the imaginary, where desire is captivated and suspended. (Eagleton 1990, 90; my emphasis)

Or, what is worse, we would be immersed in the (s)mothering feminine slime of the unconscious, the feminine here split as a defence mechanism, as it was for the fascist soldier males, between the good woman in white and the bad, red woman, and even the black water of the swamp monster/mother.

These processes can be construed in historical and environmental terms, as J. J. Bachofen (1967) did to some extent, when the mountainous and phallic sublime of the capital(ist) city violates and colonises the beautiful feminine enclosures of English agriculture and threatens them with reversion to the slimy feminine of the swamps, the fens, the unconscious, the 'primitive' and the indigenous.[16] In response to this threat, the masculine sublime and the project of modernity maintain themselves by 'managing' in neo-imperialism the constant threat of inundation posed by the feminine and pre-modern slimy. Thus, Eagleton argues at length:

> art for Hegel is not properly representational at all, but rather an intuitive presentation which expresses a vision rather than imitates an object. It incarnates a sensuous awareness of the Absolute, abolishing all contingency and so showing forth *Geist* in all its organic necessity. As with Kant's aesthetic, this sensuousness is securely non-libidinal, freed from all desire, a beauty whose otherwise disruptive material forces is defused by the spirit which informs its every part. This intimate yet idealised body, material yet miraculously undivided, wordlessly immediate yet shaped and stylised, is what Romanticism terms the symbol, and psychoanalysis the body of the mother [and mother earth]. It is therefore not surprising that Hegel finds something obscure and enigmatic about this form of materiality, something within it troublingly resistant to the translucent power of reason. Its betrays its unsettling force most obviously in the 'bad infinity' of oriental or Egyptian art, that grotesque spawning of matter ['primeval slime'?], vague, groping and boundless, which threatens in its fantastic heterogeneity to *swamp* [my emphasis] the pure spirit like some nightmarish creature of science fiction [such as *Alien* or *Swamp Thing*]. Hegel finds such sublime proliferation of matter distinctly unnerving [what Schelling and Freud called uncanny]: within his system, this shapeless feminine stuff [or slime] is redeemable only when impregnated by rational form [a Bachelor Machine for a Bachelor Birth], negated in its material presence and gathered into the inner unity of

[16] For further discussion of Bachofen, see Giblett (1996, 34–36).

> the Idea [in other words, subl(im)ated]. The oriental stage of art is, so to speak, the child smothered by a fleshly mother; in the harmonious artefacts of ancient Greece and Rome, child and mother have achieved some symmetrical unity; at the highest, Romantic stage of art, when the almost disembodied spirit yearns for freedom from its material encasement, the child is in the process of definitively overcoming the Oedipal crisis and taking leave of the mother altogether. We do not, then, remain with the aesthetic long, but climb up [out of the 'primeval slime' as it were] a stage to religion, which still presents the Absolute in terms of images; and finally, if we can stay the course, we rise to the rarefied conceptual representations of philosophy itself. (Eagleton 1990, 143–144)

And so attain to the sublime heights of capitalism, theory, and the superego in which patriarchal men can calculate themselves as independent of nature in the Kantian dynamical sublime in which 'man' exercises power over 'nature' and regards himself as superior to nature, by contrast with the mathematical sublime in which reason is more powerful than 'man' and man is inferior to reason (Kant 1952, 109, 111). They can then repudiate their connections with their mothers, their own bodies, and with mother earth, and can continue to exploit the natural environment with reasonable impunity, a highly dubious, masculinist, and self-destructive enterprise, though the triumph of rationalist and capitalist modernisation and of modernity, not least over monstrous wetlands and their monsters.

The sublime, Eagleton (1990, 213) argues later, "for Marx as for Kant is *Das Unform*: the formless or monstrous." The swamp monster and the monstrous swamp have a shape (which are similar to each other, even homologous), but they do not have a Platonic form. They do not conform to a pre-existing or transcendental pattern or ideal, but are variable and mutable. They are sublime and sublimated. The function of art, Freud (1985b, 140–141) suggested in his essay 'Creative Writers and Day-Dreaming,' is to sublimate repudiated or repressed elements of sexuality by producing socially acceptable daydreams. Sublime art certainly does not serve to imitate nature.

Indeed, in the aesthetics of the sublime, Lyotard (1989, 202) argues, "art does not imitate nature, it creates a world apart […] in which the monstrous and formless have their rights because they can be sublime."[17] Perhaps it is hardly surprising then, as Eagleton (1990, 90) notes, that

[17] See also Deleuze (1984, especially 50–52).

"Kant associates the sublime with the masculine and the military," especially in the light of the war against wetlands, the kind of war that the fascist Mussolini liked to fight because it was so winnable in the short term with modern industrial machinery. This association bears out further Zoë Sofoulis" contention that slime is the obverse of the sublime for the masculine and military sublime sublimates, or wars, on the feminine and pacific slimy by rendering it monstrous thus trying to justify its own taming of the wet wilderness through drainage and wet-'land-filling' of rubbish tips.

Monsters not only serve this function in relation to sublimation, but also serve an important, and related, function in the taxonomies of natural history. Monsters, François Jacob (1973, 27) argues, "always bear resemblances, but they are distorted resemblances which no longer correspond to the normal action of nature." In other words, monsters correspond to the pathological action of nature, but a nature pathologised by culture. Or as Michel Foucault (1970, 155) puts it, monsters form "the background noise, as it were, the endless murmur of nature." Monsters are the black noise or interference in the communication channel/canal through which human 'culture' talks to itself about non-human 'nature,' including black waters, in that endless murmur of conversation about 'nature' of which nature documentaries are only an institutionalised part. Without that noise, communication would not be possible. No doubt marsh monsters and swamp serpents serve these socially useful functions, but at the cost of psychogeopathologising the marsh and swamp themselves within patriarchal western culture.

Historically within patriarchal culture the creation of monsters points to the ongoing struggle between two orders, the old order of traditional matrifocal swamps and the new order of patriarchal natural history. With the rise of "the new biology" François Jacob (1973, 124) quotes Isidore Geoffrey Saint-Hilaire to the effect that

> monstrousness is no longer a random disorder, but another order, equally regular and equally subject to laws: it is the mixture of an old and a new order, the simultaneous presence of two states that ordinarily succeed one another.

Jacob (1973, 124) goes on to add that "all types of monstrosity are not possible" and that "monsters can be classified" in "teratology," as 'nature' more generally is categorised in phenology and taxonomy. The taxonomy of monsters is a way of mastering the monstrous. Monsters combine elements of known creatures, a head from there, a torso from elsewhere, a lion's head and a snake's body or a human head and a lion's body.[18]

References

Bachofen, J. J. 1967. *Myth, Religion and Mother Right: Selected Writings.* Translated by Ralph Manheim. Princeton, NJ: Princeton University Press.

Bauman, Zygmunt. 1973. *Culture as Praxis.* London: Routledge and Kegan Paul.

Benjamin, Walter. 1979. *One-Way Street and Other Writings.* Translated by Edmund Jephcott and Kingsley Shorter. London: NLB.

Benjamin, Walter. 1999. *Selected Writings: Volume 2, 1927–1934.* Edited by Michael Jennings, Howard Eiland and Gary Smith. Translated by Rodney Livingston and others. Cambridge, MA: The Belknap Press of Harvard University Press.

Benjamin, Walter. 2003. In *Selected Writings: Volume 4, 1938–1940*, ed. Howard Eiland and Michael Jennings. Cambridge, MA: Belknap Press of Harvard University Press.

Berman, Marshall. 1983. *All That Is Solid Melts Into Air: The Experience of Modernity.* London: Verso.

Burke, Edmund. 1958. In *A Philosophical Enquiry Into the Origin of Our Ideas of the Sublime and Beautiful*, ed. James T. Boulton. London: Routledge and Kegan Paul.

Carrouges, Michel. 1954. *Les machines célibitaires.* Paris: Arcanes.

Certeau, Michel de. 1984. *The Practice of Everyday Life.* Translated by Stephen F. Rendall. Berkeley, CA: University of California Press.

Certeau, Michel de. 1986. The Arts of Dying: Celibatory Machines. In *Heterologies: Discourse on the Other.* Translated by Brian Massumi, 156–167. Manchester: Manchester University Press.

Cixous, Hélène, and Catherine Clément. 1986. *The Newly Born Woman.* Translated by Betsy Wing. Manchester: Manchester University Press.

Collins, Margery L., and Christine Pierce. 1976. Holes and Slime: Sexism in Sartre's Psychoanalysis. In *Women and Philosophy*, ed. Carol C. Gould and Mark W. Wartofsky, 112–127. New York: Putnam.

Couture, Jean-Pierre. 2016. *Sloterdijk.* Cambridge, UK: Polity Press.

Daly, Mary. 1979. *Gyn/Ecology: The Metaethics of Radical Feminism.* London: Women's Press.

Dante, Alighieri. 1982. *The Inferno.* Translated by John Ciardi. New York: Signet Classics. First published in Italian in the fourteenth century of the Common Era.

[18] The monstrous is discussed further in Chap. 10.

Deleuze, Gilles. 1984. *Kant's Critical Philosophy: The Doctrine of the Faculties.* Translated by Hugh Tomlinson and Barbara Habberjam. London: Athlone.
Deleuze, Gilles. 1989. *Masochism: Coldness and Cruelty.* Translated by Jean McNeil. New York: Zone Books.
Deleuze, Gilles, and Felix Guattari. 1977. *Anti-Oedipus: Capitalism and Schizophrenia.* Translated by Robert Hurley, Mark Seem, and Helen R. Lane. New York: Viking.
Derrida, Jacques. 1982. White Mythology: Metaphor in the Text of Philosophy. In *Margins of Philosophy*, 207–271. Translated by Alan Bass. Brighton, Harvester.
Eagleton, Terry. 1990. *The Ideology of the Aesthetic.* Oxford: Basil Blackwell.
Estok, Simon C. 2024. *Slime: An Elemental Imaginary.* Cambridge, UK: Cambridge University Press.
Forester, C. S. 1956. *The African Queen.* Harmondsworth: Penguin. First published 1935.
Foucault, Michel. 1970. *The Order of Things: An Archaeology of the Human Sciences.* London: Tavistock.
Freud, Sigmund. 1964. *New Introductory Lectures on Psycho-Analysis, Standard Edition, Volume XXII.* London: Hogarth.
Freud, Sigmund. 1973. *Introductory Lectures on Psychoanalysis*, Penguin Freud Library, 1. Harmondsworth: Penguin.
Freud, Sigmund. 1985a. *Civilization, Society and Religion*, Penguin Freud Library, 12. Harmondsworth: Penguin.
Freud, Sigmund. 1985b. *Art and Literature: Jensen's Gradiva, Leonardo da Vinci and Other Works*, Penguin Freud Library 14. Harmondsworth: Penguin.
Giblett, Rod. 1996. *Postmodern Wetlands: Culture, History, Ecology.* Edinburgh: Edinburgh University Press.
Giblett, Rod. 2011. *People and Places of Nature and Culture.* Bristol: Intellect Books.
Giblett, Rod. 2019. *Psychoanalytic Ecology: The Talking Cure for Environmental Illness and Health.* London: Routledge.
Giblett, Rod. 2024. *Wetland Cultures: Ancient, Traditional, Contemporary.* New York: Palgrave Macmillan.
Giblett, Rod. 2025. *Waking to Nature with Thoreau and Benjamin: Or, Pscyhopolitical Ecology.* New York: Bloomsbury.
Giblett, Rod. 2026. *Wise Women for Loving Nature: Or, Femopolitical Ecology.* New York: Bloomsbury.
Grosz, Elizabeth. 1994. *Volatile Bodies: Toward a Corporeal Feminism.* St Leonards, NSW: Allen and Unwin.
Grosz, Elizabeth. 1995. *Space, Time and Perversion: The Politics of Bodies.* St Leonards, NSW: Allen and Unwin.
Han, Byung-Chul. 2017. *Psychopolitics: Neoliberalism and New Technologies of Power.* Translated by Erik Butler. London and New York: Verso.

Harvey, Robert. 1991. The Sartrean Viscous: Swamp and Source. *SubStance* 64:49–66.

Hebdige, Dick. 1987. The Impossible Object: Towards a Sociology of the Sublime. *New Formations* 1:47–76.

Hill, Christopher. 1965. *God's Englishman: Oliver Cromwell and the English Revolution*. London: Weidenfeld and Nicolson.

Irigaray, Luce. 1985. *Speculum of the Other Woman*. Translated by Gillian C. Gill. Ithaca, NY: Cornell University Press.

Irigaray, Luce. 1993. *Je, Tu, Nous: Toward a Culture of Difference*. Translated by Alison Martin. New York: Routledge.

Jacob. 1973. *The Logic of Life: A History of Heredity*. Translated by Betty E. Spillman. New York: Pantheon.

Kant, Immanuel. 1952. *The Critique of Judgement*. Translated by James Creed Meredith. Oxford: Clarendon Press. First published in German in 1790.

Kant, Immanuel. 1960. *Observations on the Feeling of the Beautiful and Sublime*. Translated by John T. Goldthwait. Berkeley: University of California Press. First published in German in 1764.

Kristeva, Julia. 1982. *Powers of Horror: An Essay on Abjection*. Translated by Leon S. Roudiez. New York: Columbia University Press.

Kristeva, Julia. 1986. About Chinese Women. In *The Kristeva Reader*, ed. Toril Moi, 138–159. Oxford: Basil Blackwell.

Laplanche, Jean, and Jean-Baptiste Pontalis. 1973. Sublimation. In *The Language of Psychoanalysis*. Translated by Donald Nicholson-Smith, 431–434. London: Hogarth.

Longinus. 1965. On the Sublime. In *Classical Literary Criticism*. Translated by T. S. Dorsch, 99–158. Harmondsworth: Penguin.

Lyotard, Jean-François. 1984. Answering the Question: What is Postmodernism? Translated by Regis Durand. In *The Postmodern Condition: A Report on Knowledge*, 71–82. Translated Geoff Bennington and Brian Massumi. Manchester: Manchester University Press.

Lyotard, Jean-François. 1989. The Sublime and the Avant-Garde. In *The Lyotard Reader*, ed. Andrew Benjamin, 196–211. Oxford: Basil Blackwell.

Lyotard, Jean François. 1993. *Libidinal Economy*. Translated by Iain Hamilton Grant. London: Athlone.

Miller, David. 1989. *Dark Eden: The Swamp in Nineteenth-Century American Culture*. Cambridge: Cambridge, UK University Press.

Muir, John. 1987. *My First Summer in the Sierra*. New York: Penguin. First published 1911.

Mumford, Lewis. 1934. *Technics and Civilization*. New York: Harcourt Brace Jovanovich.

Noordegraaf-Eelens, Liesbeth, and Willem Schinkel. 2011. The Space of Global Capitalism and its Imaginary Imperialism: An Interview with Peter Sloterdijk.

In *In Media Res: Peter Sloterdijk's Spherological Poetics of Being*, ed. Willem Schinkel and Liesbeth Noordegraaf-Eelens, 185–194. Amsterdam: Amsterdam University Press.

Poynton, Cate. 1985. *Language and Gender: Making the Difference*. Geelong: Deakin University Press.

Sartre, Jean-Paul. 1969. *Being and Nothingness: An Essay on Phenomenological Ontology*. Translated by Hazel E. Barnes. London: Methuen. First published in French in 1943.

Schaffer, Kay. 1988. *Women and the Bush: Forces of Desire in the Australian Cultural Tradition*. Cambridge: Cambridge University Press.

Simmel, Georg. 1950. The Metropolis and Mental Life. In *The Sociology of Georg Simmel*, 409–424. Translated by Kurt H. Wolff. New York: Macmillan.

Sloterdijk, Peter. 1987. *Critique of Cynical Reason*. Translated by Michael Eldred. Minneapolis, MN University of Minnesota Press. First published in German in 1983.

Sloterdijk, Peter. 2010. *Rage and Time: A Psychopolitical Investigation*. Translated by Mario Wenning. New York: Columbia University Press. First published in German in 2006.

Sloterdijk, Peter. 2011. *Spheres, Volume 1: Bubbles: Microspherology*. Translated by Wieland Hoban. Los Angeles, CA: Semiotext(e). First published in German in 1998.

Sloterdijk, Peter. 2012. The Time of the Crime of the Monstrous: On the Philosophical Justification of the Artificial. In *Sloterdijk Now*, edited by Stuart Elden, 165–181. Translated by Wieland Hoban. Cambridge, UK: Polity Press.

Sloterdijk, Peter. 2013a. *Philosophical Temperaments: From Plato to Foucault*. Translated by Thomas Dunlap. New York: Columbia University Press. First published in German in 2009.

Sloterdijk, Peter. 2013b. *You Must Change your Life: On Anthropotechnics*. Translated by Wieland Hoban. Cambridge, UK: Polity Press. First published in German in 2009.

Sloterdijk, Peter. 2014. *Spheres, Volume 2: Globes: Macrospherology*. Translated by Wieland Hoban. Los Angeles, CA: Semiotext(e). First published in German in 1999.

Sloterdijk, Peter. 2017. *The Aesthetic Imperative: Writings on Art*, edited by Peter Weibel. Translated by Karen Margolis. Cambridge, UK: Polity Press. First published in German in 2014.

Sloterdijk, Peter. 2024. *Out of the World*. Translated by Corey A. Dansereau and Gill Zimmermann. Stanford, CA: Stanford University Press. First published in German in 1993.

Sloterdijk, Peter. 2025. *The Terrible Children of Modernity: An Antigenealogical Experiment*. Translated by Oliver Berghof. New York: Columbia University Press. First published in German in 2014.

Sofoulis, Zoë. 1988. Through the Lumen: Frankenstein and the Optics of Re-Origination. Ph.D. Thesis, History of Consciousness, University of California, Santa Cruz.

Solnit, Rebecca. 2000. *Wanderlust: A History of Walking*. London: Penguin.

Solnit, Rebecca. 2005. *Hope in the Dark: The Untold History of People Power*. Edinburgh: Canongate.

Solnit, Rebecca. 2007. *Storming the Gates of Paradise: Landscapes for Politics*. Berkeley, CA: University of California Press.

Spengler, Oswald. 1932. *The Decline of the West*, Volume II. Translated by Charles Francis Atkinson. London: George Allen and Unwin.

Theweleit, Klaus. 1987. *Male Fantasies*, volume I: *Women, Floods, Bodies, History*. Translated by Stephen Conway. Cambridge, UK: Polity.

Theweleit, Klaus. 1989. *Male Fantasies, Volume II: Male Bodies: Psychoanalyzing the White Terror*. Translated by Chris Turner and Erica Carter. Cambridge, UK: Polity.

Thirsk, Joan. 1957. *English Peasant Farming: The Agrarian History of Lincolnshire from Tudor to Recent Times*. London: Routledge and Kegan Paul.

Thompson, E. P. 1968. *The Making of the English Working Class*. Harmondsworth: Penguin.

Thoreau, Henry David. 1962. In *The Journal 1837–1861, Volumes I–XIV*, ed. Bradford Torrey and Francis Allen. New York: Dover.

Thoreau, Henry David. 1980. A Walk to Wachusett. In *Natural History Essays*, 31–50. Salt Lake City, UT: Gibbs Smith.

Tremayne, Peter. 1985. *Swamp*. New York: St Martin's Press.

Van Veen, Joh. 1949. *Dredge Drain Reclaim: The Art of a Nation*. The Hague: Trio.

Virilio, Paul, and Sylvere Lotringer. 1983. *Pure War*. Translated by Mark Polizzotti, New York: Semiotext(e).

Williams, Raymond. 1985. *The Country and the City*. London: Hogarth Press.

CHAPTER 4

Mythology, Polytheism, and the Uncanny

Abstract The sublime and sublimation are always haunted and made possible by their other of slime and the uncanny, the combination of fascination and horror. The uncanny is associated with the obverse of the sublime. This chapter traces the uncanny from the work of Schelling that comes out of his philosophy of polytheism and mythology, the bases for monotheism and theology. At a time and for a world in which some sectors of the monotheistic religions are in aggressive, and sometimes violent, contention against those of other monotheistic religions, and even against other sectors within their own monotheistic religion, Schelling's discussion of the uncanny enables rethinking the relationship between mythology, and monotheistic and polytheistic religions in a culturally and politically liberatory and progressive way in the Anthropobscene. This chapter makes a plea for polytheism in a world and age riven by conflicts between and within monotheistic religions. Polytheism is living with nature in the earthly household of the ecosphere. The uncanny is the means to do it. It taps into a timely and rich vein of thinking about the divine, darkness, light, aura, religion, mythology, and the monstrous that enables rethinking the inter-relationship between all of them in environmentally and animal friendly terms.

At a time and for a world in which some sectors of the monotheistic religions are in aggressive, and sometimes violent, contention against those of

R. Giblett, *Living with Nature in the Anthropobscene*,
https://doi.org/10.1007/978-3-032-12822-5_4

other monotheistic religions, and even against other sectors within their own monotheistic religion, Friedrich Schelling's discussion of the uncanny enables rethinking the relationship between mythology and monotheistic and polytheistic religions in a culturally and politically liberatory and progressive way.[1] The monotheistic sectors generally claim special and privileged access to the truth by way of divine revelation in their scriptures and/or via their priests and the more fundamentalist ones don't acknowledge their roots in mythology or, if they do, in polytheism. By equating mythology with polytheism, Schelling prioritises polytheism over monotheism, whereas polytheism is anathema for the Judaic and Christian monotheistic religions as proclaimed in the first interdiction of the Ten Commandments, "you shall have no other gods before me" (Exodus 20: 3). Schellingian polytheism deconstructs this commandment by countermanding to the monotheistic God in dialogic reciprocity: 'there are many other gods before you.'

These aspects of Schelling's thinking about the uncanny are absent from Sigmund Freud's classic essay on the uncanny. Freud's engagement with Schelling's uncanny is sketchy, superficial, and peremptory to say the least. Freud (2003, 132) cites Schelling's discussion of the uncanny in his essay on the uncanny, but does not give a precise reference. Schelling's uncanny is relatively hidden and secret in Freud's essay, and only occasionally comes to light or out into the open in later discussions (like the uncanny itself). Schelling's uncanny is uncanny in Freud's essay. Similarly, some Anglophone Schellingians mention a source for Schelling's definitions of the uncanny, but these references are often as uncanny as Freud's reference in his essay to Schelling's uncanny and to the story that he recounts that he had found in *The Strand* magazine about a table with carved crocodiles and that he thought exemplifies the uncanny (as we will see in the following chapter).

Schelling's uncanny is not only uncannily present/absent in Freud's essay, but also completely absent from some studies by Anglophone writers on Schelling, mythology, and the unconscious. For instance, S. J. McGrath's tantalisingly entitled *The Dark Ground of the Spirit: Schelling and the Unconscious*, in which one might expect to find a discussion of the uncanny given the title and subtitle, does not mention or

[1] For a brief discussion of Schelling's philosophy (without mentioning the uncanny), see Sloterdijk (2013, 59–63). For a more extensive discussion of Schelling's philosophy (also without mentioning the uncanny), see Mules (2014).

discuss the uncanny at all (McGrath 2012). Similarly, Victor Hayes's, *Schelling's Philosophy of Mythology and Revelation*, a translated "reduction" of, and commentary on, Schelling's three lecture series on these topics in the second series of which he defines and discusses the uncanny, is in the same boat as McGrath with no mention of it (Hayes 1995). By contrast, some writers more interested primarily in Freud and his concepts of the unconscious and/or the uncanny (rather than in Schelling and his discussion of the unconscious and/or the uncanny) address the uncanny in Schelling's work.

The present chapter argues that Schelling's discussion of the uncanny taps into a timely and rich vein of thinking about the divine, darkness, light, aura, religion, mythology, and the monstrous overlooked by Freud that enables rethinking the inter-relationship between all of them in less pejorative terms than Christian theology does, including ecotheology, and in more environmentally- and animal friendly terms (as ecocultural studies does; see Giblett 2018, chapter 4). As well as locating the primary source of Schelling's definition of the uncanny cited by Freud, the chapter interweaves this bibliographic solution with a discussion of mythology and polytheism in Schelling's work and in the secondary literature. The chapter is variously: a bibliographic detective story; a contribution to the burgeoning literature on the uncanny; a contemporary political and theological intervention into monotheistic contention; and a work of environmental philosophy and activism by rethinking religion through mythology and poetics as a way of bringing being and living with the earth into closer affinity with nature.

4.1 Schelling's Uncanny and Freudian Psychoanalysis

In his essay on 'the Uncanny' first published in 1919, Freud (2003, 132) cites Schelling's definition: "uncanny is what one calls everything that was meant to remain secret and hidden and has come out into the open." Rather than translating "*hervorgetreten*" (Freud 1955, 256[2]) as "come out into the open," the *Standard Edition* of Freud's works translates the German as "come to light" (Freud 1955, 224). "*Hervorgetreten*" is

[2] The German text of the extract from Daniel Sander's *Worterbuch der Deutschen Sprache* cited in Freud's essay is helpfully included as an appendix in *The Standard Edition*, with the '2, 2, 649' cryptic reference to Schelling (Freud 1955, 252–256).

usually translated as "emerged," to which the later 2003 Penguin translation seems closer than the older *Standard Edition* translation, which implies enlightenment both as an individual process and philosophical moment and movement. Terry Castle (1995, 7) bases her study of the invention of the uncanny as the counter to enlightenment in eighteenth-century culture on the flimsy foundation of this dubious translation in the *Standard Edition*.

Schelling emphasises more the religious and mythological aspects of the uncanny than the philosophical and psychological ones. Freud does not cite Schelling directly, but cites his definition from a secondary text in the form of Daniel Sanders's *Wörterbuch der Deutschen Sprache* (*Dictionary of the German Language*) published in 1860 that, in turn, cites "Schelling 2.2, 649" (more of this cryptic reference later, which the translation in the *Standard Edition* omits). The precise source of Schelling's definition remains hidden in Freud's essay and does not come out into the open. Freud later appropriates Schelling's definition for his own definition of the uncanny in terms of psychological repression, whereas Schelling's definition emphasises pantheistic mythology. Schelling's philosophy is not about repression, but release into the possibility of a new mythology.[3]

The hidden and secret aspects of the uncanny relate particularly to Schelling's second mention of the uncanny that Freud (2003, 132) cites from Sanders: "to veil the divine and surround it with an aura of the uncanny." Schelling alludes here to the holy of holies in the Jewish temple that is veiled. Once a year the high priest is permitted to part the veil and enter the holy of holies to commune with the divine. As a result, the divine is veiled with "an aura of the uncanny." Schelling is using this example as a metaphor for the absolute, or the "thing in itself." Freud does not discuss these aspects of Schelling as cited by Sanders and they remain hidden and do not come out into the open in Freud's essay with this one exception. Aura and the uncanny are qualities that Walter Benjamin (2015, 64, 69, 72) found in early photographs, including in a photograph of Schelling taken in *c.*1850. Benjamin seems to have been unaware of Schelling's writings on aura and the uncanny, though he would have probably been aware of Schelling's post-Kantian idealism. As with Freud, Schelling's uncanny is uncannily present/absent in Benjamin's work.[4]

[3] I am grateful to Warwick Mules for this point.

[4] I am grateful to Warwick Mules for these points.

Sloterdijk (2013, 359) suggested "Freud knew something about" the uncanny and specified that

> what Sigmund Freud knew about the uncanny was merely an indication, tainted by psychologizing, of the fundamental pervertibility of the primary spatial difference between the inner-own and the outer-foreign: before all psychology there is the experience that the inner can occasionally appear foreign, while the outer can seem like the own. (Sloterdijk 2014, 197–198)

The process of what Sloterdijk (2014, 176, 178) calls "sanctifying the interior and demonising the surroundings" creates the uncanny when what he also calls "the excluded potentially return from the monstrous, that is to say the evil-infested environment." In the story Freud recounts the monstrous crocodile returns virtually from the demonised and evil-infested environment of a swamp to the sanctified interior of a London house via the vehicle and vector of a table with carved crocodile legs (as we will see in the following chapter).

Freud (2003, 123) generally secularises, internalises, and psychologises the uncanny and divests it of divine, though not of its monstrous, connotations in his own definitions of the uncanny at the beginning of his essay as belonging "to the realm of the frightening, of what evokes fear and dread." Of course, the divine can be frightening and evoke fear and dread, but Freud does not say so. On the following page, Freud more explicitly defines the uncanny as "that species of the frightening that goes back to what was once well-known and had long been familiar" (Freud 2003, 124). The divine may go back to what was once well-known and had long been familiar, but at a cultural and historical level before the fall from grace, rather than at an individual one, especially before the return to grace. The divine for Schelling remains veiled and surrounded with an aura of the uncanny. The divine is precisely that which is *not* well-known and familiar. The divine is mediated by revelation. Revelation for Schelling (2008, 100) is "something mediated through earlier processes, never [...] something *immediate*, first, original." By contrast, mythology is immediate, first, original.

Freud implicitly builds into his definition of the uncanny the tried and true concept and activity of repression. Freud (2003, 134) later explicitly evokes repression by using Schelling's definition, though for Schelling the uncanny is about the absolute shining imperfectly through the real, about hope, not repression. Sixteen pages later in the 2003 Penguin translation of Freud's essay after first citing Schelling's definition cited by Sanders, Freud (2003, 148) states that the uncanny is:

> actually nothing new and strange, but something that was long familiar to the psyche and was estranged from it only through being repressed. The link with repression now illuminates Schelling's definition of the uncanny as 'something that should have remained hidden and has some into the open.'

The uncanny Schellingian uncanny returns in Freud's uncanny albeit in the mistaken terms of repression, but Freud's other citation of Schelling's uncanny in terms of the veiled and auratic divine remains buried and hidden in Freud's essay in the form of Sanders's citation of Schelling and the cryptic reference, and never comes into the open and to light again. Freud's appropriation of Schelling's uncanny serves a repressive theory of the ego and blocks the possibility of thinking of *unheimlich* as hope, rather than repression, as coming into the open, not coming into the light.

Schelling gets short shrift from other writers on the uncanny too, more so than he does from Freud. In her recent "genealogy of the uncanny," Anneleen Masschelein (2011, 2 and 47) acknowledges that "crucial for the uncanny is the concept of repression and the return of the repressed," as it was for Freud referring to Schelling mistakenly in this regard. Yet she does not acknowledge or cite, as Freud does, Schelling's definition of the uncanny. Schelling is largely written out of her study as signalled in the subtitle of *The Freudian Uncanny in Late Twentieth-Century Theory*. Even his surname is uncannily located out of alphabetical order in the index to her book after the entry headed "Scientific" (Masschelein 2011, 227).

Freud for Masschelein is the Promethean "culture hero" who stole the concept of the uncanny from the gods and brought it down to mere mortals, whereas in fact he was the borrower of the concept from others, including Schelling and his writings on the gods. Masschelein credits Freud with raising:

> the phenomenon and the word '*unheimlich*' to the status of a concept in the foundational essay "The Uncanny" [...] Freud's essay [...] remains the primary focus of attraction in the continuing fascination with the uncanny in culture and theory alike [...] This brings us to the central thesis of this book, namely that *the Freudian uncanny is a late-twentieth century theoretical concept.* (Masschelein 2011, 3–4; her emphases)

Masschelein goes on to mention Schelling in passing in this context, but her genealogy of the uncanny does not extend beyond Freud and his essay of the early twentieth century into discussing Schelling's work on the uncanny in the nineteenth century in which the phenomenon and the word '*unheimlich*' had already been raised to the status of a concept. If not, how else could Sanders and Freud cite Schelling's definition and the latter use his definition in discussing repression? Freud has no paternal rights of primogeniture when it comes to the concept and word 'uncanny,' certainly not to the "discovery and creation of the concept in 1919," as Masschelein (2011, 4) also puts it, nor does Schelling for that matter who traces the genealogy of the phenomenon back to pre-history and pre-mythology. If Freud discovered the concept, he could not have created it.

Alix Strachey is the midwife at this strange, cross-century, and cross-cultural birth as she first translated Freud's essay '*Das Unheimlich*' into English as "'The 'Uncanny'" (and *unheimlich* as uncanny) for Freud's *Collected Papers* first published in 1925, which was then "considerably modified" for her husband's James Strachey's *Standard Edition of the Complete Psychological Works of Sigmund Freud*. An editor's note introduces Freud's essay and states that

> the first section of the present paper, with its lengthy quotation from a German dictionary, raises special difficulties for the translator. It is to be hoped that readers will not allow themselves to be discouraged by this preliminary obstacle, for the paper is full of interesting and important material, and travels far beyond merely linguistic topics. (Freud 1919, 218)

Unheimlich remains untranslated in the extract from Sanders in the *Standard Edition*, including the citation from Schelling (Freud 1955, 224). A footnote for the first appearance of the translation in the body of the text states that "the German word, translated throughout this paper by the English 'uncanny', is '*unheimlich*', literally 'unhomely.' The English term is not, of course, an exact equivalent of the German one" (Freud 1955, 219).

Masschelein (2011, 3) asserts archly that

> a theory of the uncanny before the twentieth century can only resort to the occurrence of the word or to descriptions of the phenomenon in literary texts and artistic sources [...] There was no theoretical or philosophical discourse [on the uncanny] before the twentieth century.

Yet Schelling developed a theory of the uncanny in the nineteenth century in his philosophical texts (as the present chapter will demonstrate). Masschelein does not discuss Schelling's work and test her assertion (let alone substantiate it). In an endnote Masschelein, like others (as we will see below), refers to Schelling's *Philosophy of Mythology* as the source of his definition without citing a precise reference for it in Schelling's oeuvre (Masschelein 2011, 171 n.22). Schelling's uncanny is hidden in her study and never comes into the open and to light.

Despite the important status of Freud's essay and the attraction it deserves and still enjoys, it has a marginal place in discussions of the Freudian canon and psychoanalysis. This is despite Masschelein claiming that, "at the end of the twentieth century, [Freud's] rather short treatise had outgrown "its marginal position in the Freudian canon" and is now a central text for Freudian aesthetics" (Masschelein 2011, 3, citing David Ellison). The uncanny is arguably anti-aesthetic as it counters the aesthetics of the sublime, picturesque, and beautiful (as we saw in the previous chapter). The uncanny assaults the sense of smell, whereas aesthetics addresses the senses of sight and hearing (as do the sublime, picturesque, and beautiful). Freud's essay has also arguably not outgrown its marginal position in the Freudian canon and Freudian psychoanalysis. Over the last half a century, Freud's essay on the uncanny has been, and still is, marginalised in the Freudian canon by theorists of Freudian psychoanalysis. The uncanny is interestingly and symptomatically absent from the classic vocabulary or dictionary of psychoanalytic terms (Laplanche and Pontalis 1973; in French, 1967) and from a relatively recent account of the development and context of Freud's thought (Makari 2008).

Similarly, two recent biographies of Freud also attest to the marginal position of his essay on the uncanny in the study of the Freudian canon and Freudian psychoanalysis at the beginning of the twentieth-first century. The most recent intellectual and cultural biography of Freud devotes only half a page out of its 500 pages and a footnote in a chapter on "*Lumière sombres*" ("Dark Enlightenment") to a discussion of the uncanny with the German *unheimlich* translated into French as "*l'inquiétante étrangeté*" and as the "*l'inquiétante familier*" following previous translations of Freud's essay into French (Roudinesco 2014, 269, 269 n.2, and 544). The French terms in cognate English of "the strange disquietude" and "the disquietingly familiar" are translated as "the uncanny" in the English translation (Roudinesco 2016, 215, 468 n.2 and 548). Sloterdijk (2011, 88) considers that "the French have tortured themselves translating Freud with their *étrangeté inquiétante*, whereas the connotations of the German largely disappear in the sonority of the English word *uncanny*." *Unheimlich* is untranslatable and so rates an entry in *Dictionary of Untranslatables* (Cassin 2014, 432) that insists on the uncanny bringing "what is hidden into the light," rather than into the open, and so reproduces uncritically the enlightenment metaphor.

The uncanny fares even worse in Joel Whitebook's recent intellectual biography of Freud in which it is not discussed and Freud's essay on it is not even listed in the bibliography (Whitebook 2017). The uncanny tends to be repressed in the secondary psychoanalytic literature. By contrast, the uncanny received a full-length book study in cultural and literary studies (see Royle 2003).

4.2 Schelling's Uncanny and Anglophone Schellingians

Unlike Freud, some other writers on Schelling ostensibly go to a primary source in Schelling's later work and writings for the source of his definition of the uncanny. For instance, Jason Wirth (2003, 150), a leader in Anglophone Schellingian studies, in his first book-length study of the philosopher, *Conspiracy of Life*, introduces Schelling's definition of the uncanny by saying that he gave it in "the *Philosophy of Mythology* lectures." However, like Masschelein, this book does not appear in Wirth's bibliography of Schelling's works (267), nor it does not exist in English as such (as Wirth, like Masschelein, seems to suggest or imply that it does by giving it an English title). In the body of the text Wirth then cites the following translation of Schelling's definition: "all that which should have remained in secret [*im Geheimmis*], in concealment and latency, but which has nevertheless stepped forward." Wirth (2003, 254, n.31) gives the following endnote here for this citation: "I have used Edward Allen Beach's translation of this passage in his *The Potencies of the God(s): Schelling's Philosophy of Mythology*, (Albany: State University of New York, 1994), 228." The source of Schelling's definition in Schelling's work remains hidden in Wirth's work and does not come to light in it, whereas it does in Beach's study when he translates and refers to Schelling (Beach 1994, 289, n.72). Beach also briefly mentions that Schelling's discussion of the uncanny "later struck Sigmund Freud as particularly acute" and refers to his essay (Beach 1994, 229 and 289, n.74.).

Anthony Vidler (1992, 14) in the introduction to his study of *The Architectural Uncanny* first cites Freud and his later elaboration of Schelling's definition without acknowledging Schelling. Vidler elides Freud's reference to "Schelling's definition" by using ellipses. Schelling remains hidden until Vidler (1992, 26; his ellipses), like Wirth, later cites Sanders citing Schelling and uses the following translation of Schelling's definition: "*Unheimlich* is the name for everything that ought to have remained [...] secret and hidden but has come to light," not into the open. Vidler (1992, 230, n.21) at this point refers to Sanders' *Dictionary* in an endnote. Vidler (1992, 26) goes on to state in the body of the text that "Schelling's felicitous 'definition' of the uncanny had in fact been extracted by Sanders from the *Philosophie der Mythologie* of 1835, Schelling's late attempt to synthesize the history of religion with the anthropology of cults." Unlike Freud, Vidler also explicitly goes on to refer to Schelling's work when he cites Schelling's definition again on the following page in a slightly different (hitherto unpublished) translation by Eric Randolf Miller: "all things are called uncanny which have remained secret, hidden, latent, but which have come to light," again, not into the open.

Vidler (1992, 27) also cites the context for Schelling's definition of the uncanny in which Schelling refers to "the dark and obscure power of that uncanny principle which dominated earlier religions." Darkness for Schelling has no pejorative overtones (as we will see later). He is not valorising later religions over earlier ones, enlightened ones over benighted ones, light over darkness. The uncanny is the process of the dark and obscure coming into the open. As the source of Schelling's definition of the uncanny and enunciation of the uncanny principle, Vidler (1992, 27; 231, n.22) refers in the following note to a two-volume edition of *Philosophie der Mythologie* published in Darmstadt in 1966 by the Scholarly Book Society. The source of Schelling's definition hidden in obscurity and twilight in the work of Sanders, Freud, Masschelein, and Wirth comes out in the open in Beach's and Vidler's work. As in Freud's essay, Schelling's definition is uncannily present/absent in Vidler's study as it was initially secret, hidden, and latent in ellipses and later comes to light in a footnote. Schelling's definition of the uncanny is uncanny indeed.

Like Wirth, Teresa Fenichel (2015, 206–207) in her 2015 PhD thesis at Boston College entitled, *Uncanny Belonging: Schelling, Freud and the Vertigo of Freedom*, cites Beach's translation of Schelling's definition of the uncanny. Unlike Wirth and Vidler, however, she introduces Schelling's definition by saying that he gave it in his "lectures on mythology and revelation." This is a grab bag term for Schelling's "last philosophy" comprising lectures on the philosophy of mythology and revelation, including his "historical-critical introduction to the philosophy of mythology."

Beginning in the late 1950s, continuing with his PhD at Columbia University in 1970 and culminating in book publication in 1995, Victor Hayes translated and published a reduction of, and commentary on, Schelling's three series of lectures on the philosophy of mythology and revelation, including the historical-critical introduction. However, Hayes does not include any translations of Schelling's definitions and discussions of the uncanny in the edition of his reduction published in book form in 1995. As a student and teacher of religion and philosophy (and not literature and mythology, unlike Schelling who was a student and teacher of all four, hence his status as a pioneer of transdisciplinary studies) and as an ordained Christian minister, Hayes approaches Schelling's "last philosophy" largely from a theological point of view and completely hides or obscures the uncanny as an avatar of mythology and polytheism (or paganism, "i.e., the non-Christian religions" (Hayes 1995, 8)). Schelling's uncanny remains hidden in the unreduced portions of Schelling's lectures on the philosophy of mythology and revelation and does not come to light in Hayes's edition, nor does the context of Schelling's uncanny in his discussion of Homer come out into the open for Hayes.

Beach and Vidler acknowledge that Schelling's definition and mentions of the uncanny invariably occur in the context of his discussion of Homer's epic poems, or more precisely Homeric epic poetry. For Schelling, 'Homer,' or Homeric poetry, is a product of mythology defined by Schelling (2008, 9) as Greek "tales, legends and stories, which in general go beyond historical time," both into pre-historical time and into what he refers to as *Dichtung*, translated as "poesy." Without Greek mythology, there would be no Homeric poetry. Homer did not write mythology; mythology wrote Homer. Indeed, Schelling (1989, 52) argues in *The Philosophy of Art* that "mythology and Homer are one and the same." For Schelling, religion is also a product of mythology. Without mythology, there would be no religion. Religion ignores its mythological roots at its peril and arrogates to itself a role and function as sole access to truth it does not warrant and has no basis for, certainly no scriptural basis.

Scripture is steeped in mythology. The ostensibly divinely revealed monotheistic Judeo-Christian scriptures are saturated with mythology (and so with polytheism, despite the best efforts of the first of the Ten

Commandments to keep them at bay). Mythology for Schelling (2008, 9; his emphases) in the *Historical-Critical Introduction to the Philosophy of Mythology* is "polytheism" and "in general the *system of the gods.*" Mythology is neither the system of one god, so nor is religion. Monotheism comes out of polytheism (and not vice versa). Before mythology lies what Schelling (2008, 17) calls "the dark foundry, the first forging place of mythology." This is the dark, obscure, secret, and veiled aspect of the uncanny that comes to light and out in the open in the other aspect of the uncanny. These two aspects of the uncanny come together in mythology. The uncanny is precisely the link and process between these two aspects brought together in mythology.

Mythology and polytheism for Schelling have truth, are a way to truth and even are the truth. Schelling (2008, 8) states up front that "there is indeed truth in mythology." Schelling (2008, 23) restates and refines his initial statement that "truth is in mythology, but not in mythology as such." Truth is in polytheism, in difference, not in sameness. Schelling (2008, 149; his emphasis) concludes that "there is truth in mythology *as such*," because mythology is polytheism, or more precisely what Schelling repeatedly calls "successive polytheism." Mythology for Schelling (2008, 145) is precisely "essentially successive polytheism."

Schelling (2008, 148–149; his emphasis) goes on to explain what he means by "successive polytheism":

> polytheism, considered in the entirety of its successive moments, is the way to truth and to this extent truth itself. One could conclude from this: in this way the final mythology, uniting all moments must be the true religion [...] The truth in mythology is initially and especially a *religious* one.

Conversely, the truth in religion is only later and especially a *mythological* one. All religion contains within it an "untruth." It is shot through with difference, i.e. potential for polytheism, a tendency that leads to law and prohibition. There is no truth in religion without prior mythological truth—prior historically, ontologically, and epistemologically. It is a commonplace of theology that it negates and sublates mythology; by equating mythology with polytheism, Schelling desublates and positivises theology and monotheism into polytheism.

Monotheism considered in the singularity of one moment and not in the entirety of the successive moments of polytheism prior to and over it is not the way to truth, is not the true religion and not primarily the truth.

In the contemporary context of the early twentieth-century, Schelling's prioritising of polytheism over and before monotheism may enable the sectors of the monotheistic religions currently in often violent contention to co-exist. The polytheisms have done so for millennia with mutual respect and cross-fertilisation. The 'three religions of China' (Taoism, Buddhism, and Confucianism) are a case in point (Fung Loy Kok Institute of Taoism® 2008, 2012[5]).

Schelling (2008, 89; his emphases) earlier considered mythology "as history of the gods, thus the actual mythology, was only able to be produced in life itself; it *had* to be something *lived* and *experienced*." The truth (of mythology and religion) is something lived and experienced. For Schelling (2008, 150) later "the content of mythology is not an *abstract*-religious one," but a concrete-religious one. The uncanny is lived and experienced truth as Freud (2003, 144, 151) shows in his account of getting lost in the red-light district of an Italian town and in his recount of the story of the table with carved crocodiles (as we will see in Chap. 5). The truth is not something believed in in the abstract, a dead, disembodied truth, but a concrete, living, embodied truth.

The way to truth for Schelling lies in nature, or what he calls "the sensible world." Nature for Schelling is the first object of knowledge and the site in which the search for God first takes place. "The immediate object of human knowledge," Schelling (2008, 56) argues, "remains nature, or the sensible world; God is only the dark, vague goal that is strived for and that is first sought in nature." Against the view that nature is God, or pantheism, Schelling (2000, 31) in *The Ages of the World* counters that "nature is not God." Seeking God in nature is polytheistic because mythology is polytheistic, not pantheistic, and truth lies in mythology (and in religion insofar as it is mythological, and not in religion insofar as it neglects or denies its mythological roots and arrogates to itself privileged access to the truth in its scriptures and/or via its priests). Mythology is the way to the truth in life (*contra* Jesus in John 14: 6 ("I am the way, the truth and the life"), or Jesus, or the Messiah, is another name, like Homer, for mythology). Mythology for Schelling (2008, 155) "indisputably has the closest affinity with nature." Theology indisputably does not. Theology has a fraught and problematic non-affinity with nature (see Giblett 2018).

[5] Grateful acknowledgement is made to the Board of Directors of the Fung Loy Kok Institute of Taoism for permission to cite these two texts.

Rethinking religion through mythology and poetics is a way of bringing being and living with the earth into closer affinity with nature.[6]

Similarly, art has a fraught and problematic non-affinity with nature. Nature for Schelling (1989, 88) is sublime, but "not sublime in itself" as the sublime is "actually found in the subject" (90). Theodor Adorno (1984, 91) bemoaned the fact that Schelling stood at the beginning of "an almost exclusive concern" in aesthetics with art and that he did not carry on with Kant's "most perspicuous investigation" into "the beautiful in nature." Yet Schelling's view of the sublime is Kantian. Art and nature were mutually exclusive categories for Schelling. Mythology, not art, for him had the closest affinity with nature because mythology arose out of nature whereas art arose out of culture.

Along similar lines to similar ends, Warwick Mules (2014) argues for a demythification of mythological thinking (though it is difficult to see in Schellingian terms how a demythified thinking could have affinity with nature). By contrast, I am arguing (here and elsewhere) for a detheologisation, desublimation, and remythification of thinking about, and regaining affinity with, nature, especially via swamps and marshes, swamp serpents, and marsh monsters, including crocodiles (as we will see in Chap. 5). One way to do so leads through Schelling's uncanny and his philosophy of mythology, and through the uncanny labyrinth of his bibliography.

Matt Ffytche (2013, 160) in *The Foundation of the Unconscious: Schelling, Freud and the Birth of the Modern Psyche*, cites Schelling's definition and footnotes '*Philosophie der Mythologie*, 649,' the same page number as Sanders cites. Ffytche later suggests that "Freud may have looked up the source context of the quotation in Schelling's lectures." If he had done so, he may have found a kindred spirit interested in literature and mythology. Ffytche's (2013, 302) bibliography lists "*Philosophie der Mythologie*, in *Sämmliche Schriften*, Abt. 2, vol. II," the same reference as Sanders cites (2.2, 649). Sanders's cryptic reference is decoded, the crypt opened, and the precise source for Sanders's citation of Schelling's definition comes to the light. Schelling's definition of the uncanny is uncanny indeed.

Elsewhere, in the *Historical-Critical Introduction to the Philosophy of Mythology*, Schelling mentions the uncanny on several occasions that contribute to an understanding of mythology and polytheism, especially the

[6] For further discussion of these topics, see Giblett (2011, 2018).

association between the divine and the monstrous. He links "religious terror" and "uncanny feelings" (Schelling 2008, 14), refers to "the dark and uncanny power of the belief in the gods" (45) and "the thoughtless dread before something uncanny" (54), and associates the uncanny with the excessive, even monstrous (65). Yet the uncanny does not rate an index entry in this translation and remains relatively hidden in it (239). Although Freud does not make the link that Schelling does between the uncanny, the divine, thoughtless dread, religious terror, and the monstrous, he might have done so when he recounted the story in *The Strand* magazine that he read involving a table with carved crocodiles (Freud 2003, 151; Moberly 1917, 1991). Crocodiles are a vehicle and vector for the uncanny, and for the divine and/or monstrous before they were separated out (as we will see in Chap. 5).

The crocodiles in the story are also associated with the obscure and darkness, qualities of the uncanny for Schelling. In the *Historical-Critical Introduction to the Philosophy of Mythology* Schelling (2008, 63) argues that when God created light and darkness, and "the antithesis" between them as "the first creation," "God calls the light good without naming the darkness evil." For Schelling (2000, 4) in *The Ages of the World* "everything begins in darkness." Everything was good, including darkness, when God created it and God loudly declared it to be so in "repeated assurance," as Schelling puts it. Some Christian theologians, such as Norman Habel and Claus Westermann, follow suit in their readings of the biblical creation story in Genesis 1 (see Giblett 2018, chapter 1). The darkness associated with the uncanny is not necessarily evil, nor is the monstrous, though the monstrous in Judeo-Christian terms is evil. Cain, the murderous brother of Abel, created evil monsters in the form of Nephilim, or giants, who were "warriors of old" (Genesis 6: 4) and in the form of the marsh monsters Grendel and his mother in the interpolation of new verses in Genesis in the Old English *Beowulf* (see Giblett 2018, chapter 2). The monstrous is not evil in non-Christian terms. The uncanny is one way of thinking darkness and the monstrous as not evil.

The monstrous is frightening as it goes back to what was once well-known and had long been familiar. It goes back to the first act of fratricide, back to the past. The monstrous crocodiles in the story that Freud condenses is a figure for the colonial repressed that returns, and to which the protagonists return, by way of the colonial artefact of the table with carved crocodiles on it. The monstrous crocodiles are a vehicle and vector for the uncanny in which the repressed not only returns, but also to which one returns. The table with carved crocodiles on it returns one of the protagonists in the story to the swamp in which a friend was taken by a crocodile. The uncanny not only entails the return *of* the repressed, but also the return *to* the repressed. Schelling's uncanny is a rich, dark brew which yields nourishment for the body and soul for those who value spirituality and materiality. It provides a progressive understanding of religion in the contemporary world of the contending sectors of monotheistic religions that might lead to a new appreciation for polytheism.

A theory of the uncanny during and before the twentieth century can resort (and does so in the present chapter) to Schelling's definition and discussions in his philosophical texts. In the twentieth-first century the Schellingian uncanny is a theoretical concept that deserves to become a focus of attraction in the continuing fascination with the uncanny in culture and theory, in politics and practices alike, especially in the contemporary context of contention between some sectors of the monotheistic religions, and between them, and Schelling's prioritising of polytheism not merely before, but also over monotheism.

4.3 Monotheism and Polytheisms

The monotheistic religions of Judaism, Christianity, and Islam are not "the three monotheisms" (Sloterdijk 2009) as that would make them polytheistic, condemned by all three religions. Sloterdijk (2009, 19, 47) also refers more precisely to "the trio of monotheistic religions" and "the three monotheistic syntheses" without addressing the slippage to them from "the three monotheisms" in the subtitle of his book.[7] There is only one God in and for the three monotheistic religions. Hence there is monotheism. 'Monotheism*s*' is a contradiction in terms. Although God is (also) called variously Yahweh, Jehovah, and Allah in and by the three monotheistic religions and their languages, s/he/it is one God known by different names, in different languages, by different peoples.

God, of course, is an English word derived from the German '*Gott*' so it is impossible to escape the prison-house of language(s). This being could be called 'x' and made equivalent to, and used as shorthand for, all the other names, but that would be taking the name of the Lord in vain in contravention of the third of the Judeo-Christian Ten Commandments

[7] Sloterdijk (2016, 4, 23) later tried to avoid "as far as possible" using "the term 'monotheism'" after an unproductive debate with theologians "over the correct use of the term."

(Exodus 20: 7). Besides, 'x' has already been taken by another god so that would create problems of trademark ownership and copyright. I will continue to call this being 'God' with one name in line with a recent authorised translation of *The Qur'an* (2016). Yahweh, Jehovah, and Allah are not separate gods, otherwise the three so-called monotheistic religions would in fact be polytheistic. This point of logic is not lost on this recent translation of *The Qur'an* into English (*The Qur'an* 2016). If translators of *The Qur'an* into English had transliterated the Arabic الله as 'Allah' and not translated it as 'God,' readers of English might not understand that this is the monotheistic God. It could also be confused with "*Allah*, a chief god," which implies other gods, and who mediated other deities for some (*The Qur'an* 2016, xii), and so is polytheistic on both counts. The polytheistic "*Allah*, a chief god," is not the monotheistic God-chief.

Polytheism constitutes all three monotheistic religions, as Schelling argued. Polytheism makes monotheism possible. Before there were believers and 'disbelievers' in God, there were only polytheists. *The Qur'an* (2016) condemns polytheists or idolators on many occasions as 'disbelievers' as they do not believe (only) in (the one and only, single and solitary) God. As *The Qur'an* puts it early on in sura 5: 73, "there is only one God" (*The Qur'an* 2016, 121) and reiterates in sura 6: 19 that "he is only one God" (*Qur'an* 2016, 131). After the monotheistic God was constituted and instituted, for monotheism there were no polytheists. *The Qur'an* traces this development in which polytheists disappear from view to be replaced by disbelievers. The monotheistic religions of Judaism and Christianity constitute and institute themselves by explicitly interdicting polytheists (or idolators and pagans) as with the first of the ten commandments. Islam followed suit.

Polytheists in their own terms are believers in their multiple and multifarious gods and goddesses; they are not necessarily unbelievers in God. Being an unbeliever in God is not an option as it is not recognised as a valid category by the three monotheistic religions. There is no opt-out category and clause in monotheism for polytheists from being either believers or 'disbelievers' in God. One is either a believer or a disbeliever on either side of a mutually exclusive binary opposition giving rise to the dualism of good versus evil in Manichaeism, a Christian heresy and supposedly defunct religion (according to the *Routledge Encyclopedia of Philosophy*), but still alive and well for evangelicals and fundamentalists, and their fellow-travellers who portray immigrants as evil. Believers are good, disbelievers are evil. For monotheism, one cannot be both a believer in God and a believer in (other) gods and goddesses.

The three monotheistic religions of Judaism, Christianity, and Islam should be able to co-exist as they all believe in (the one and only, single and same) God. *The Qur'an* embraces the founders (Noah, Abraham, Moses, and Jesus) of the other two monotheistic religions as "messengers of firm resolve" from God, along with Muhammad "the Prophet" (*The Qur'an* sura 46: 35; 2016, 507n). In *The Qur'an* "the Prophet" is instructed in sura 2: 139 to ask Jews and Christians, "how can you argue with us about God when he is our Lord and your Lord?" (*The Qur'an* 2016, 22). All the points of difference between the monotheistic religions are secondary and epi-theological compared to their shared belief in God. For the individual Jew, Muslim or Christian, the individual followers of the other two religions are fellow-believers; they are not 'disbelievers,' a point lost on some believers of all three religions. If they were disbelievers, they would be condemned, which they are by some believers, or worse in Islam in a misinterpretation of sura 9: 5 (*The Qur'an* 2016, xxiii, 188). In so doing they are taking God's judgement into their own hands and arrogating to themselves the divine function of "the power to decide," as *The Qur'an* (2016) says on many occasions, either in the present with disasters and destruction, or in the future with the final judgement and eternal damnation in the fires of hell. God also exercised mercy towards disbelievers who repented and became believers in God, as *The Qur'an* (2016) also says on many occasions.

4.4 Taoism

By contrast, the three (polytheistic) 'religions' or teachings of China (Taoism, Buddhism, and Confucianism) co-exist with each other, other polytheistic religions (e.g. Hinduism), and the three monotheistic religions. According to the Fung Loy Kok Institute of Taoism® (FLK), Taoism has "integrated" Buddhism and Confucianism (FLK 2008, vi–vii). In addition, "Taoism is polytheistic, which means that there are myriad deities, immortals, lords, spirits, and so on in the Taoist pantheon" (FLK 2008, 113). No mention is made here of 'religion.' Is Taoism a religion? Of course, it all depends on how you define 'religion' (discussed further in the following chapter). The Fung Loy Kok Institute of Taoism® addresses this question straight up and gives a direct answer to it by stating that "in many Eastern traditions, religion is seen more as a school of teaching than as a devotional faith with strict doctrines as is often seen in the West" (FLK 2008, vi). Taoism is a "school of teaching" that does not exclude other schools of teaching (or 'religions'). Taoism is a way of the way (or Tao). Taoism follows "the Great Way (the Tao)" (FLK 2008, 19). Similarly, Taoist deities are not gods in the sense of being separate beings, but are embodiments of virtues or expressions of the body and mind. They are not without but within. For instance, Taoism embraces Goon Yam/Guan Yin, the Bodhisattva, or Buddhist goddess, of compassion. Taoism is a way of cultivating bodily, spiritual, and environ-mental health in the Taoist body and with Taoist ecology. The Taoist body and Taoist ecology combine in the Taoist body of the earth in which the body and the earth are not separate entities, but embody and figure each other with the body as land and vice versa in Taoist ecology within the body and without.[8]

Sloterdijk (2020, xi) claims in his book on "Eurotaoism" that "even Taoism can't help us anymore." My book on Taoism claims that it is an ongoing way of helping ourselves and each other in line with the Buddhist virtue of compassion and the Confucian virtues of piety, harmony, and so on (Giblett 2023). Taoism can help, especially with the biospiritual practices of *tai chi*, what Sloterdijk (2020, 28) calls "being-towards-stillness within-movement," and meditation, being-towards-movement-within stillness. They are more helpful than being towards death and are akin to being born (as we will see in Chap. 10). Other Taoist arts of health, such as performing rituals and chanting sutras, also help. Unbeknown apparently to Sloterdijk, tens of thousands of Europeans, North and Central Americans, Australians, and New Zealanders practice Taoist Tai Chi® and other Taoist internal arts of health and find them immensely helpful as recounted and documented in their 'testimonials' and health recovery stories.[9] Taoism is still helping.

Taoism also helps to think, figure, and act with the body and the universe differently from western ways. Master Moy Lin-Shin, the founder of Taoist Tai Chi® and the Fung Loy Kok Institute of Taoism®, wrote in his one and only letter to practitioner-members in 1994 that

> we are not learning taijiquan, or the Tai Chi Boxing set; we are learning Tai Chi, the Grand Ultimate. The Yin and Yang Tai Chi symbol contains the Universe and explains its functioning. Similarly, the 108 moves of the Taoist

[8] For further discussion of these topics, see Giblett (2023).

[9] For further discussion of these topics, see Giblett (2008b).

> Tai Chi® arts set reflect the Universe in our bodies, and can enhance the working of this Universe within.

The famous *yin* and *yang* symbols together make-up *tai chi*, "the grand ultimate" that contains and explains the Universe without and within, unlike the tradition in western culture in which the universe is the container and the body is contained in it and the disembodied mind explains the universe. For Taoism there is one universe, not two universes, which would be a contradiction in terms (like monotheisms). The universe within and without are united, not separate. The 108 moves of the Taoist Tai Chi® 'set' embody and enhance the functions of the Universe within and the Universe without, both intertwined in *yin* and *yang*. "The grand ultimate" (*tai chi*) of *yin* and *yang* of the universe within and without intertwine and complement each other—they are not binary opposites. *Yin* and *yang* comprising *tai chi* are the logo for the Taoist Tai Chi® Society.

The Taoist body and ecology of *yin* and *yang* deconstruct and decolonise the western spheres within spheres, including the biosphere and ecosphere, into the bioregional places of home-habitats of the living earth for the body (and the mind/spirit, as we will see in the final chapter). Taoist ecology of the body of the earth embodies and enhances the functioning of the Universe without. The Taoist body performed in the rhythmic internal arts of health is also unlike the western body as machine, or battlefield, or the sporting body imprisoned in the time-machine, or the surface of static inscription for fashion, sport, and the law, and like the grotesque body of the lower strata and the monstrous body of the slimy depths.[10] Taoist ecology of the body of the earth is unlike the western earth as machine, or source of resources, or surface of inscription for roads and towns, and like(s) the grotesque lower earthly strata and the monstrous slimy earthly depths. The *Tao Te Ching*, the primary text of Taoism, states in Chap. 8 that "the Tao loves the watery places loathed by men," such as the slimy and sacred swamps loved by Thoreau (as we will see in Chap. 5).

[10] For discussion of the grotesque body of the lower strata, the monstrous body of the slimy depths, the modern western medical body as machine, the Fascist body as war-machine, the sporting body imprisoned in the time machine, the body as surface of inscription, the hypermodern body as battlefield, and the cyborg body-machine of the civilian-soldier, see Giblett (2008a).

4.5 Water

Water exemplifies the soft and yielding power of the Tao, or the Way, and the Taoist body (expressed especially in *tai chi*) that overwhelms and undermines the hard and rigid body of the machine. For the Taoist the body and the earth are consanguineous and conterminous because they are united and enlivened by the same energy flow, or *Ch'i/qi*. *Ch'i/qi* or matter/energy is central to a Taoist view of the body and of the earth, the body of the earth. *Ch'i/qi* flows through the earth like energy through the body. In Chinese cosmology the body and the earth are both made of the same living, vital breath, or *Ch'i/qi*. *Ch'i/qi* is linked to water which is both the life-blood of the earth and the major component of the human body. The same *Ch'i/qi* is pumped and flows through our veins and arteries, airways and lungs, as flows through the rest of nature. Yet Taoism is not alone in figuring the body as earth, and its fluids as water. In pre-modern western medicine the body is figured as earth and the earth as body, whereas in modern western medicine the body is figured as machine. Leonardo da Vinci (1452–1519 CE) figured the body in terms of both. He is on the cusp of this transition. The diabolic genius of da Vinci was not only to propound this view of the body as earth or landscape, but also to combine it with the view of the body as machine. The sacred genius of Lao Tzu and Guan Tzu (as we will see shortly) was to propound the view of the body as earth with its flows of water, the lifeblood of the earth.

The *Tao Te ching* contains two explicit and important mentions of water, the first in verse 8:

> The best way to live
> is to be like water
> For water benefits all things
> and goes against none of them
> It provides for all people
> and even cleanses those places
> a man is loath to go
> In this way it is just like the Tao (Lao Tzu 2001, 21)

One of the places where "a man is loath to go" is wetlands. Yet wetlands are a place where water already is. How can water cleanse a place where water already is? That is part of the paradox of water—and the Tao. It is already in places where a man is loath to go and it cleanses those places. Or perhaps more precisely, the plants of the place absorb the nutrients and filter the water to make it cleansing. This water provides for all people and benefits all things. It thus gives an exemplar of the way to live. By going low, by going to low places, by taking the lower part, by going into the nether lands and into the grotesque lower bodily and earthly regions, water not only goes to places men are loath to go but also cleanses, provides and benefits all people and all things (see Giblett 1996, 2008a).

The other mention of water in the *Tao Te Ching* in verse 78 also highlights the paradoxical power of water—and the Tao:

> Nothing in the world
> is as soft and yielding as water
> Yet for attacking the hard and strong
> none can triumph so easily
> It is weak, yet none can equal it
> It is soft, yet none can damage it
> It is yielding, yet none can wear it away
> Everyone knows that the soft overcomes the hard
> and the yielding triumphs over the rigid (Lao Tzu 2001, 91)

The rock wears away from water flowing over it, but the water never wears. Water is the wearing agent. Water exemplifies the soft and yielding power of the Tao, and of the Taoist body (expressed especially in *tai chi*), that overwhelms and undermines the hard and rigid body of the machine (expressed especially in the Fascist body and body-building; see Giblett 2008a).

The slow, graceful movements of *tai chi* are like a gently flowing river; the stillness in movement and the movement in stillness of *tai chi* are like the still waters of a wooded wetland reflecting the heavens above; the moving meditation of *tai chi* is like a deep ocean pool full of waving and weaving mysteries; the power and grace of *tai chi* is like a serpentine stream curving and straightening through the body of the earth; the strength and flexibility of the Taoist body in *tai chi* is like a continental river with deep currents carrying mighty ships and with shallow billabongs giving birth, nutrition, and shelter to small fish; the internal circulation of energy in the body in the practice of *tai chi* is like a underground stream coursing through the cavities and waters of the lower earthly stratum cleansing and purifying them; the rising of energy from the feet, through the pelvis, and up the spine to the top of the head in *tai chi* is like a bubbling spring welling up from deep underground to bathe the brain, the "palace of the mud ball," in cerebro-spinal fluid; the internal alchemy of *tai chi* rejuvenates like a delta or lagoon giving birth to new life, the most fertile places on the planet; the generating and storing of energy, the accumulation of *chi*, in *tai chi* is like an estuary to which all good things flow and which expels bad things; and *yin* and *yang* complementing each other are like the black waters of slimy swamps and the white waters of mountain streams, both part of the body of the earth—the Tao is like water.

In his notes to verse 78 of *Tao Te ching* Jonathan Star elaborates that

> although each element of nature can be likened to the Absolute, water has always been the element most endeared by the Taoist and the one closest to the nature of Tao. Water not only represents humility, gentleness, and a mind in perfect repose, but it is the element that gives birth to all life. Witness this brilliant passage on water from the *Kuan-tzu* [*Guanzi*] – a Taoist work of the fourth century B.C.E:
>
> water is the blood of the Earth; it flows through its muscles and veins… It is accumulated in Heaven and Earth and stored up in the various things of the world. It comes forth in metal and stone and is concentrated in living creatures. Therefore it is said that water is something with a spirit. Being accumulated in plants and trees, their stems gain orderly progression, their flowers obtain proper number, and their fruit gain proper measure. Water gathers in jade, and the nine virtues appear. It congeals to form man […] that is its most refined essence […] Nothing is produced without it. Only he who knows how to rely on its principles can act correctly. (Lao Tzu 2001, 269)

For the Taoist, the earth is a body and water is the life-blood of the earth.

Moreover, for the Taoist, as Allan (1997, 4) puts it, water "is the model for philosophical ideas about the nature of the cosmos." It is also the model for personal conduct and for life itself. The Classical Chinese word, *shui*, conventionally translated as 'water' is thus, Allan (1997, 32) argues, "a broader category than the English 'water'." Confucius (cited by Allan 1997, 24) said that "water, which extends everywhere and gives everything life without acting (*wuwei*) is like virtue (*de*) […] Its bubbling up, never running dry, is like the way (*dao* [*Tao*])." Water is thus the model for three of the central concepts of Confucianism and Taoism: the *dao* or *tao*, the way, itself; *de*, virtue; and *wuwei*, nonaction. For Allan (1997, 31):

> water served as a root metaphor in the formulation of abstract concepts that were the basis of the system of social and ethical values.

It also served in the same capacity in relation to environmental and corporeal values, and to the earth and the body.

The body and the earth form the body of the earth, what Sarah Rossbach (1984, 8) calls "a sacred metabolic system" of energy flows in living beings, rather than the secular metabolic exchange of dead matter in modern industrial capitalism (see Giblett 2008a, chapter 1). Whereas the former is enacted in the cultivation and circulation of energy in the living body of the earth, the latter culminates in the production of the commodity and in the exchange of dead matter and labour for money, and of money for dead matter. And whereas the latter gives rise to nature as "man's [*sic*] inorganic body" from which 'he' is alienated and on which 'he' works in the production of commodities, the former is intimately related to nature as humans' body with which humans are united and (with) which humans work (see Giblett 2008a, 2011).

For the Taoist the body and the earth are consanguineous and conterminous because they are united and enlivened by the same energy flow, or *Ch'i/qi*. *Ch'i/qi* or matter/energy is central to a Taoist view of the body and of the earth, the body of the earth. *Ch'i/qi* for Wei-Ming (1989, 76) is "the blood and breath of the vital force that underlies all beings." The *Guanzi (Chuang Tzu)* calls water "the blood and breath of the earth" (cited by Kuriyama 1999, 50). *Ch'i/qi*, or vitality as Allan (1997, 60) translates it, is "closely associated with breath – and mist, the breath of the earth [...] and in later theory it is said to be that which everything is made of." For Cheng (1994, 117) "the vital breaths simultaneously animate both the being of the universe and the being of man [*sic*]." One breath, two beings: human being and the being of the universe are distinct and not merged into one whole but breathe the same breath.

Ch'i/qi contrasts with some major western concepts, such as matter. *Ch'i/qi*, according to Graham (cited by Allan 1997, 88) occupies:

> The place in Chinese cosmology occupied by matter in ours. The basic metaphor behind the word matter is of timber (Latin *materia*), inert and cut up and to be assembled by a carpenter; *qi*, on the other hand, is in the first place breath, alternating between motion and stillness.

This alternation between motion and stillness is expressed in *tai chi*. *Tai chi* involves what Marcel Mauss (1992, 469) calls "dancing at rest;" it is also what he calls "a technique of active rest." The techniques of active rest, such as *tai chi*, can be contrasted with the techniques of passive speed of modern communication and transportation technologies (see Giblett 2008b).

In Maussian terms the narrator in Julia Kristeva's (1992, 169) novel *The Samurai* describes *tai chi* as "a kind of tortoiselike eurythmics." One character goes on to elaborate in Maussian terms that in *tai chi*:

> You dance with your arteries and veins […] – it's not a matter of arms and legs. Your blood surges forward, draws back, and then the time comes when your body's transformed. It doesn't disappear, but the rhythm of your blood is in harmony with shapes you imagine in space – your body itself is all space, what's inside and outside are brought together and redistributed, magnified almost to infinity, anatomised to infinity.

Tai chi is an external and internal dance of bodily limbs, organs, and fluids choreographed to the imagined shapes and actions of animal movements: stork cools wings, snake creeps down, golden cock stands on one leg, and so on. Every activity of the Taoist body, and not only *tai chi*, is, in Kohn's (1991, 229) terms, "patterned on the rhythm of the world at large."

In Chinese cosmology the body and the earth are both made of the same living, vital breath (see Cheng 1986, 362–363). In western cosmology the body and the earth are both made of the same inert and desiccated matter. Matter is dry whereas breath is wet. Blood is wet too, and so *ch'i/qi* is linked to it, and both are linked to water, which is the life-blood of the earth and the major component of the human body. Water in the *Guanzi* is "the blood and *qi*-breath of the earth" (cited by Allan 1997, 123). Conversely, *chi/qi* is the water and breath of the body. Consequently for Wei-Ming (1989, 78) "we are consanguineous with nature." We human beings share the same blood, and breath, of *chi* as nature. The same *chi* is pumped and flows through our veins and arteries, airways and lungs, as flows through the rest of nature. The Taoist body and ecology are alive and well; Taoism is still helping us human beings. Taoist ecology is an exemplary psychopolitical ecology for cultivating environ-mental health and for curing environ-mental illness.[11]

[11] For further discussion of the Taoist body, Taoist ecology, and Taoist Tai Chi®, see Giblett (2023).

References

Adorno, Theodor. 1984. The Beauty of Nature. In *Aesthetic Theory*, ed. Gretel Adorno and Rolf Tiedemann, 91–115. Translated by Christian Lenhardt. London: Routledge and Kegan Paul.

Allan, Sarah. 1997. *The Way of Water and Sprouts of Virtue*. Albany, NY: State University of New York Press.

Beach, Edward. 1994. *The Potencies of the God(s): Schelling's Philosophy of Mythology*. Albany, NY: State University of New York.

Benjamin, Walter. 2015. *On Photography*. Edited and translated by Esther Leslie. London: Reaktion Books.

Cassin, Barbara. 2014. Das Unheimliche. In *Dictionary of Untranslatables: A Philosophical Lexicon*, ed. Barbara Cassin, 432. Princeton, NJ: Princeton University Press.

Castle, Terry. 1995. *The Female Thermometer: Eighteenth-Century Culture and the Invention of the Uncanny*. Oxford: Oxford University Press.

Cheng, C.-y. 1986. On the Environmental Ethics of the *Tao* and the *Ch'i*. *Environmental Ethics* 8 (4): 351–370.

Cheng, F. 1994. *Empty and Full: The Language of Chinese Painting*. Boston: Shambhala.

Fenichel, Teresa. 2015. Uncanny Belonging: Schelling, Freud and the Vertigo of Freedom. PhD thesis, Boston College.

Ffytche, Matt. 2013. *The Foundation of the Unconscious: Schelling, Freud and the Birth of the Modern Psyche*. Cambridge, UK: Cambridge University Press.

Freud, Sigmund. 1955. The 'Uncanny.' In *The Standard Edition of the Complete Psychological Works of Sigmund Freud, Volume XVII (1917–1919): An Infantile Neurosis and Other Works*, 217–256. Translated by Alix and James Strachey. London: Hogarth. First published in German in 1919.

Freud, Sigmund. 2003. *The Uncanny*. Translated by David Mclintock. London: Penguin. First published in German in 1919.

Fung Loy Kok Institute of Taoism®. 2008. *A Path of Dual Cultivation: Teachings of the Fung Loy Kok Institute of Taoism*. Toronto: Fung Loy Kok Institute of Taoism.

Fung Loy Kok Institute of Taoism®. 2012. *Ten Thousand Buddhas Sutra*. Translated by the Fung Loy Kok Institute of Taoism. Toronto: Fung Loy Kok Institute of Taoism.

Giblett, Rod. 1996. *Postmodern Wetlands: Culture, History, Ecology*. Edinburgh: Edinburgh University Press.

Giblett, Rod. 2008a. *The Body of Nature and Culture*. Basingstoke: Palgrave Macmillan.

Giblett, Rod. 2008b. *Health Recovery: The Taoist Tai Chi™ Way*. London: Shepheard Walwyn.

Giblett, Rod. 2011. *People and Places of Nature and Culture*. Bristol: Intellect Books.

Giblett, Rod. 2018. *Environmental Humanities and Theologies: Ecoculture, Literature and the Bible*. London: Routledge.

Giblett, Rod. 2023. *The Way of Taoism: For Bodily, Spiritual, and Environmental Health*. London: Transnational Press.

Hayes, Victor. 1995. *Schelling's Philosophy of Mythology and Revelation: Three of Seven Books Translated and Reduced with a General Introduction*. Armidale: Australian Association for the Study of Religions.

Kohn, Livia. 1991. Taoist Visions of the Body. *Journal of Chinese Philosophy* 18:227–252.

Kristeva, Julia. 1992. *The Samurai: A Novel*. Translated by Barbara Bray. New York: Columbia University Press. First published in French in 1990.

Kuriyama, Shigehisa. 1999. *The Expressiveness of the Body and the Divergence of Greek and Chinese Medicine*. New York: Zone Books.

Lao Tzu. 2001. *Tao te Ching: The Definitive Edition*. Translated by Jonathan Star. New York: Tarcher.

Laplanche, Jean, and Jean-Baptiste Pontalis. 1973. *The Language of Psychoanalysis*. Translated by Donald Nicholson-Smith. London: Hogarth. First published in French in 1967.

Makari, George. 2008. *Revolution in Mind: The Creation of Psychoanalysis*. New York: Harper.

Masschelein, Anneleen. 2011. *The Unconcept: The Freudian Uncanny in Late-Twentieth Century Theory*. Albany, NY: State Univrsity of New York Press.

Mauss, Marcel. 1992. Techniques of the Body. In *Incorporations*, ed. Jonathan Crary and Sanford Kwinter, 455–477. New York: Zone.

McGrath, Sean. 2012. *The Dark Ground of the Spirit: Schelling and the Unconscious*. East Sussex: Routledge.

Moberly, L. G. 1917. Inexplicable. *The Strand Magazine* 54 (324): 572–581.

Moberly, L. G. 1991. Inexplicable. In *Strange tales from The Strand*, ed. Jack Adrian, 183–195. Oxford: Oxford University Press. First published 1917.

Mules, Warwick. 2014. *With Nature: Nature Philosophy as Poetics Through Schelling, Heidegger, Benjamin and Nancy*. Bristol: Intellect Books.

Rossbach, Sarah. 1984. *Feng Shui: Ancient Chinese Wisdom of Arranging a Harmonious Living Environment*. London: Rider.

Roudinesco, Élisabeth. 2014. *Sigmund Freud en son Temps et dans le Nôtre*. Paris: Seuil.

Roudinesco, Élisabeth. 2016. *Freud: In His Time and Ours*. Translated by C. Porter. Cambridge, MA: Harvard University Press.

Royle, Nick. 2003. *The Uncanny*. New York: Routledge.

Schelling, F. W. J. 1989. *The Philosophy of Art*. Edited and translated by Douglas Stott. Minneapolis, MN: University of Minnesota Press.

Schelling, F. W. J. 2000. *The Ages of the World (Fragment) from the Handwritten Remains: Third Version (c. 1815)*. Translated by Jason Wirth. Albany, NY: State University of New York Press.

Schelling, F. W. J. 2008. *Historical-Critical Introduction to the Philosophy of Mythology*. Translated by Mason Richey and Markus Zisselsberger. Albany, NY: State University of New York Press.

Sloterdijk, Peter. 2009. *God's Zeal: The Battle of the Three Monotheisms*. Translated by Wieland Hoban. Cambridge, UK: Polity. First published in German in 2007.

Sloterdijk, Peter, with Hans-Jürgen Heinrichs. 2011. *Neither Sun nor Death*. Translated by Steve Corcoran. Los Angeles, CA: Semiotext(e). First published in German in 2001.

Sloterdijk, Peter. 2013. *Philosophical Temperaments: From Plato to Foucault*. Translated by Thomas Dunlap. New York: Columbia University Press. First published in German in 2009.

Sloterdijk, Peter. 2014. *Spheres, Volume 2: Globes: Macrospherology*. Translated by Wieland Hoban. Los Angeles, CA: Semiotext(e). First published in German in 1999.

Sloterdijk, Peter. 2016. *In the Shadow of Sinai: A Footnote on the Origins and Changing Forms of Total Membership*. Translated by Wieland Hoban. Cambridge, UK: Polity Press. First published in German in 2013.

Sloterdijk, Peter. 2020. *Infinite Mobilization: Towards a Critique of Political Kinetics*. Translated by Sandra Berjan. Cambridge, UK: Polity Press. First published in German in 1989.

The Qur'an: English Translation with Parallel Arabic Text. 2016. Translated by M. A. S. Abdel Hasseen. Oxford: Oxford University Press.

Vidler, Anthony. 1992. *The Architectural Uncanny: Essays in the Modern Unhomely*. Cambridge, MA: MIT Press.

Wei-Ming, Tu. 1989. The Continuity of Being: Chinese Visions of Nature. In *Nature in Asian Traditions of Thought: Essays in Environmental Philosophy*, ed. J. Baird Callicott and Roger Ames, 67–78. Albany, NY: State University of New York Press.

Wirth, Jason. 2003. *The Conspiracy of Life: Meditations on Schelling and his Time*. Albany, NY: State University of New York Press.

Whitebook, J. 2017. *Freud: An Intellectual Biography*. Cambridge, UK: Cambridge University Press.

CHAPTER 5

The Uncanny, Crocodiles, and Religion

Abstract Freud took up the uncanny in his essay of 1919 on the topic, arguably his greatest, most fruitful, and enduring contribution to the study of culture and nature. This chapter identifies its eco-cultural aspects that he neglected, especially his reading of what he calls "a naïve, but extraordinarily uncanny story" that he "came across" about crocodiles and a table with carved crocodile legs. This chapter argues that the uncanny is the return *to* the repressed that brings into the open in the present what is in the past and long forgotten and opens a pathway to hope in and for the future. It also critiques the gender politics, ecoculture and psycho- and femo- political ecology of a late twentieth-century Australian eco-feminist's recount of her encounter with a monstrous crocodile in an uncanny swamp. It concludes by asking the question what is religion? and answers that religion is the uncanny without the boundaries of mere reason. The uncanny does not have reason and has absented itself from it. The uncanny swamp for Thoreau is religion in his definition as "where our love is." Religion is love for places and places that love, such as swamps for Thoreau in which he found hope.

R. Giblett, *Living with Nature in the Anthropobscene*,
https://doi.org/10.1007/978-3-032-12822-5_5

5.1 Freud's Uncanny and the Monstrous

The concept/metaphor of the uncanny is arguably the greatest and most fruitful contribution of Sigmund Freud to the study of culture and nature. Freud developed the uncanny in his essay of this title first published in 1919 in the psychoanalytic journal *Imago*. It was first translated into English by Alix Strachey in 1925 for Freud's *Collected Papers*. This translation was then "considerably modified" for her husband's James's *Standard Edition of the Complete Psychological Works of Sigmund Freud* published in 1955 and republished by Penguin in their 'Penguin Freud Library' in 1985. A new translation was commissioned and published by Penguin early this century (Freud 2003).

While a whole swag of Freud's concepts and ideas (such as the Oedipus complex, penis envy, and so on) are critiqued and problematised, or poohed-poohed and dismissed as the years go by, the uncanny endures for a century as a useful tool in the toolbox of cultural criticism, critical theory and political and philosophical critique. Freud (1955, 219; cf. 2003, 124) defines the uncanny as "that class of the frightening which leads back to what is known of old and long familiar." This is specifically, though Freud does not mention it, what Jeanne Randolph (2001b, 97) calls "the mother-infant embrace," both *in-utero* and *ex-utero*. For Freud (1955, 244, 2003, 151) the uncanny is literally *unheimlich*, unhomely, but also homely. The uncanny is what Freud (1957) called "a primal word" with "antithetical meanings" and contradictory feelings which he found associated in the minds of adult males with female genitalia and in his own mind with the first home of individual human life in the mother's womb. The uncanny for Freud entails a return of the repressed.

The return of the repressed occurs here and elsewhere for Freud as he does not refer to the mother—his or anybody else's—or to her body, or specifically her womb. The maternal body is repressed in and by Freud in his essay on the uncanny as Randolph (2001a, 184, 2001b, 97) points out. Freud's uncanny is uncanny in which the repressed maternal body returns. Freud (2003, 151) relates that "it often happens that neurotic men state that to them there is something uncanny about the female genital organs. But what they find uncanny ['unhomely'] is actually the entrance to man's old 'home,' the place where everyone once lived." Freud is, of course, referring to the womb, but seems a bit squeamish about doing so explicitly. 'Man's' old home is the womb, whereas the old men's home is (the entrance to) the tomb. The womb is more than a place (where action occurs, or a site where action takes place; it is not a passive receptacle). It is where the processes of life begin and are nurtured. It is also where the first bond occurs *in-utero* and the mother's body is where the first bond occurs *ex-utero.*

Life *in-utero* is a time and space when, as Freud (1955, 235; cf. 2003, 143) describes it euphemistically, "the ego had not yet marked itself off sharply from the external world and from other people." Freud (1955, 235) goes on to remark that "these factors are partly responsible for the impression of uncanniness." Both Freud's euphemisms of "the external world" and "other people" for the mother's body are factors partly responsible for the impression of uncanniness in Freud's essay on the uncanny. Life *in-utero* and *ex-utero* is a time and space when the ego had not set itself against anything, including the external world, and from other people, such as the mother. The mother herself, as Randolph (2001b, 98) puts it following Melanie Klein, is "the infant's 'external world'" and the "other people" Freud vaguely refers to is the mother. "Freud's weakness" for Sloterdijk (2011, 217) is "the unwillingness to think the mother"—not to mention thinking the great mother of the swamps (Bachofen 1967). Freud could have started thinking the great mother of the swamps when he gave a perfunctory and dismissive reading of a story about crocodiles living in a swamp (as we will see later in the present chapter),

The ego was a part and parcel of the inside world *in-utero* and the uncanny other, the mother, and of the outside world *ex-utero.* Randolph (2001a, 186) argues that 'The Uncanny' Freud misses "the whole point about mothering." Twenty years later in his London notes of 1938 Freud (1975, 299) gets closer to the point about mothering by mimicking the child: "'the breast is a part of me, I am the breast'." Sloterdijk (2011, 95, 293, 301) critiques in his "groundwork for a negative genecology" this deification of "'the mother's breast' [as] that alleged first 'object' (note the singular)." He counters later that "here are not yet traces of subject- or object-like aspects in the actual situation." There are what Kristeva (1982, 1–2) calls abjects between and prior to subject and object. All 'objects' are initially conflated with one another and equated with the mother, or parts of her, especially her breasts, as she/they are the primary 'objects' for the young infant. Melanie Klein, as Randolph (2001b, 98) puts it, "gave the mother her due," as Margaret Mahler did too in her work on psycho-symbiosis and Jessica Benjamin in her work on "the bonds of love" (discussed in the final chapter).

The home place and processes, or perhaps more precisely the 'unhome,' of the uncanny is not only the mother's womb, but also the wetland, the first home of life on the planet in the earth's womb and tomb from both of which new life springs and is nurtured (see Giblett 2018, chapter 1). The wetland is the uncanny place and process *par excellence*. It is also the home, or unhome, of alligators and crocodiles, the 'king' of the tropical wetland, the obverse of the temperate dry land, and the archetypal swamp monsters *par excellence*. In some cultures, the alligator and crocodile are not only monstrous, but also divine. In Judeo-Christian theology, the monstrous and divine are divided, and the monstrous demonised; in the uncanny, the divine and monstrous are re-united.

The uncanny is applicable not only to the b(l)ack waters of the swamp, but also to the dark underworld of the city which for Freud were an object, or more precisely abject, of horror and fascination. Freud (2003, 144) relates in 'The Uncanny' how he was:

> Strolling one hot summer afternoon through the empty and to me unfamiliar streets of a small Italian town. I found myself in a district about whose character I could not long remain in doubt. Only heavily made-up women were to be seen at the window of the little houses, and I hastily left the narrow street at the next turning. However, after wandering about for some time without asking the way, I suddenly found myself back in the same street, where my presence began to attract attention. Once more I hurried away, only to return there again by a different route. I was now seized by a feeling that I can only describe as uncanny, and I was glad to find my way back to the piazza that I had recently left and refrain from any further voyages of discovery.

Freud's lost object is himself that he repeatedly finds in the wrong place, or more precisely, the place to which he unintentionally returns. This return enacts an unconscious desire. The uncanny is a return *to* the repressed.

For some nineteenth-century writers on the city its underside was exclusively an object of horror and for some twenty-first century writers the slum is an object of fascination. The dark underside of the city is often figured in terms of colonial places and spaces, such as the jungle and the swamp (as we will see in Chap. 9). These writers are 'placist' in that they ascribe characteristics to a (human-made) place in the city, such as the slums, that were previously ascribed to a place not made with human hands, or made with ancient human hands (jungle, abyss, nether-land, swamps, etc.), such as darkness, decomposition and disease. As monsters are demonised, so are monstrous places. These places are theologised and moralised as places of evil, the monstrous and the demonic. They need to be detheologised, desublimated, and remythified in much more environmentally friendly terms of, for example, thinking the city and the swamp as the body.

The preceding chapter brought into the open the uncanniness of Schelling's uncanny in Freud's essay, including some aspects of the uncanny ignored by Freud (as a lapsed or secular Jew), such as its birth in mythology and polytheism, and its association with the divine and monstrous (ignored by both Freud and Schelling). The present chapter continues to make the plea for valuing polytheism over monotheism, the monotheistic religions and those monotheistic sectors contending against each other, often aggressively and violently, in the world today with often disastrous and devastating consequences for people and places with loss of life and ruination of cities and country-sides.

The present chapter argues for the detheologisation, desublimation, and remythification of thinking religion through mythology and for polytheistic poetics as a way of bringing being and living with the earth into closer affinity with nature. It prefers polytheology, the dialogic study of multiple, multifarious and polysemic religions, gods and goddesses with a multivocal play of voices traced in performances, to monotheology, the study of one monologic religion and the capital 'G' God with a monopoly on a single, univocal truth inscribed in a written scripture. It also considers monsters and the monstrous (neglected by Schelling and Freud) as and at the intersection of the divine and the human that became separated and the monstrous demonised in the development of monotheism, and later rationalism.

For Freud, the alligator and the crocodile portrayed as orally sadistic monsters are vehicles and vectors of the uncanny. This combination gives rise to what could be called 'the monstrous uncanny' in which the fascinating and horrific are projected onto, and embodied in, an orally sadistic monster. The uncanny counters the aesthetics of the sublime and the beautiful (as we saw in Chap. 2). Whereas the latter privilege the distancing sense of sight, the uncanny engages the sense of smell which is much more up close and personal, though not as immediate as touch and taste. Smell is often commented upon in encounters, real or imagined, with an alligator or crocodile.

The monstrous uncanny not only engages the olfactory, but also the oral and tactile. The uncanny associated with smell (as Freud did) is a sublimation to some extent of the uncanny associated with taste and touch. The uncanny is evoked by what is not seen, which could not only be what is smelt, but also what is tasted and touched, both of which are involved when being eaten. The fear of being killed and eaten, and the desire not to be, are understandable, and certainly characterise the human side of the relationship with alligators and crocodiles. It also presumably characterises the other side too. The relationship between eater and eaten is ultimately non-reciprocal: only one being gets to eat and the other to be eaten. When humans venture into the habitat of alligator or crocodile they can be prey and "being prey" as Val Plumwood describes it is a terrifying experience, but also instructive as it was for her (as we will see later in the present chapter). Being prey highlights the non-reciprocal nature of the relationship. Besides consuming alligators and crocodiles as meat, humans also consume their habitat by destroying wetlands in a colonising and non-reciprocal relationship.

Alligators and crocodiles have been living on the earth for 200 million years, much longer than any other currently surviving, similar-sized genus of the animal kingdom with the exception of sharks. They are truly a 'blast from the past.' As such, they are vehicles and vectors for the uncanny. Freud (1985, 340) defined the uncanny as "that class of the frightening which leads back to what is known of old and long familiar." The uncanny is not only a return to the past, but also in quasi-Freudian terms 'a return *to* the repressed,' including the colonial repressed. What was repressed for the nineteenth-century, petit-bourgeois and Viennese Freud was invariably sexual in nature.

More generally in patriarchal culture the repressed is what is corporeal, visceral, maternal and monstrous. These can include the sexual, oral and anal, all of which can be, and were, associated with the colonial. In Freud's 'The Uncanny' the crocodile emerges as a figure for the British colonial repressed to which he returns via the vehicle and vector of the artefacts of colonialism which bear the traces of other, alien or exotic places and peoples. The repressed does not have a fixed content or function but changes historically and varies culturally, though the crocodile and swamp as sites of the colonial unconscious have been with us for some time as we will see. Crocodiles for the ancient Egyptians were sacred, as they are for Australian Aboriginal peoples.

How the sacred becomes the monstrous, and even demonic, is a vexed topic. For David Quammen in a reading of the biblical Leviathan in Job 41 the monster is both created by God and subdued by Him in order to instil humility in His people (Quammen 2004, 13). The monstrous and divine have been split off from each other and the divine constitutes itself and its hold over its believers by claiming to have created the monster, and, in turn, by subduing it. This splitting occurs in and as the moment of the institution of history. In pre-history the monstrous and divine are coterminous and mutually defining, and the uncanny is the name of this imbrication. Indeed, the monstrously sacred/demonic is perhaps an apt definition of the uncanny. Just as the sublime functions as a secular theology in which the sublime stands in for God in a culture for which God is dead (as we saw in Chap. 2), so the uncanny operates as a kind of secular demonology with the alligator and crocodile as Satan and the swamp as hell.

According to *The New Bible Dictionary* (Douglas 1970, 729), Leviathan is usually considered to be "some form of aquatic monster." In Psalm 44: 26 "it is clearly of the sea and is generally thought to be the whale," though in the passage from Job 41 "most scholars agree that the creature is a crocodile." The previous chapter of Job (40: 15-22) describes another monster translated as "Behemoth" in the Authorised (or King James) Version. In the Revised Standard Version "Behemoth" is footnoted as "the hippopotamus." The same version footnotes "Leviathan" in the following chapter as "the crocodile." *The New Bible Dictionary* (Douglas 1970, 138), concurs that "Behemoth" in Job is "the hippopotamus." Yet the *New English Bible* (cited by Kelly, 2006, 67 and 246) has "crocodile" for "Behemoth" in Job 40: 15 (this translation has "whale" for "Leviathan" in Job 41: 1). In this translation of the Bible the crocodile is both "the chief of beasts" and "the chief of God's beasts" (Kelly 2006, 67). The crocodile is the king of the wet jungle, of the tropical wetland, the first, and best, work of God on the second day of creation (see Giblett 2018, chapter 1).

Dragons, by contrast, may be the worst work of God, or at least the worst creation coming out of primeval chaos. In Psalm 74: 13-14 Leviathan is associated with "the dragons of the waters." Perhaps Phillip Parker (2015, 132) had this verse in mind when he described the most advanced Viking longships as leviathans. Their longships were not only dragon-prowed, but dragons themselves. Dragons can be aquatic creatures, but they can also be aerial, infernal, and terrestrial creatures (see Giblett 2018, chapter 3). In Psalm 74 Trudinger (2001, 35) comments that "great importance is given to Earth [...]; as cosmos, it is a place of harmony and order [...] The water-beings do not enjoy such a happy status. They are portrayed as God's opponents, suffer horrible injury and a degrading end." Dragons and crocodiles are no exception. God is a dryland God, not a wetland God. He is a God of dryland agriculture and pastoralism. He is opposed to wetlands, their beings, and paludiculture.

Leviathan has a much more ambiguous role and function in the Old Testament as a chaos monster and as a playmate than in the New Testament as the Devil or Satan (see Beal 2002, 80). Whereas the dragon or Leviathan of the Old Testament is not associated with the Devil or Satan, in the last book of the New Testament the dragon or Leviathan is identified with the Devil or Satan (Revelation 9: 17 and 19; see Giblett 2018, chapter 2). In the Old Testament, as Beal (2002, 81) comments, "neither Leviathan nor the sea monster nor any other monster is ever identified with the Devil or Satan." In the last book of the New Testament, however, the dragon or Leviathan are identified with the Devil or Satan, and they are locked in the battle of "good versus evil, God versus Devil" (Beal 2002, 81).

5.2 Monstrous Crocodiles

Perhaps no animal has been more deified/demonised than the crocodile, the 'monarch of the marsh' and the 'king of beasts' of the tropical swamp. The alligatorian and the crocodilian have been repressed for a long time, at least since Freud's time, and it still persists. For Vollmar (1972, ix) "crocodiles, alligators, and caimans both horrify and fascinate," a shorthand definition of the uncanny. Freud (1985, 339) defined the uncanny as "what is frightening – what arouses dread and horror" and I have defined the uncanny as what is both horrifying and fascinating (Giblett 1996). Vollmar (1972, ix) suggests that "lurid travellers' tales of evil reptiles lying loglike in tropical mud, ready to snatch and devour the unwary human, linger in the memory." Freud (1985, 367, 2003, 151) developed the uncanny from reading one such tale, L. G. Moberly's 'Inexplicable,' published in *The Strand Magazine* in 1917.

In his reading of Moberly's story Freud downplays the role of real crocodiles and ignores the swamp as the place *par excellence* of the uncanny. If I were Freud, I would psychoanalyse some long lost and repressed memories to do with his father and the phallus as the crocodile must be some sort of 'phallic symbol' in Freud's lexicon of symbols, and to do with his mother and the swamp as a maternal place. Yet rather than psychoanalysing Freud's psychopathology, I want to analyse the psychogeopathology that portrays the alligator and the crocodile as an orally sadistic monster, to engage in the talking cure of a psychopolitical ecology that would regard them and the swamp in less demonic and more sacral terms, and to promote environ-mental health that would mean that these psychogeopathological symptoms did not arise in the first place (see Giblett 2019).

Freud, perhaps in typical fashion, gives a three or four sentence summary of the story he "came across" in a magazine. He does not give a reference other than saying that it appeared in "a number of the English *Strand Magazine*" (Freud 2003, 151). Some time ago I 'came across' two precise references to the story in Royle's *The Uncanny*, one to the original publication in *The Strand Magazine* and the other to a reprint in an anthology of stories from *The Strand Magazine* (Royle 2003, 140–141, n.3; Moberly 1917, 1991). Rather than concentrating on the slips and mistakes in Freud's retelling as he himself would do, I want to consider the gaps and absences, the symptomatic lacunae, of Freud's reading in order to reinstate crocodiles living in a swamp as a vehicle and vector of the uncanny and disinvest their construction as orally sadistic monsters.

In 'The Uncanny' Freud (2003, 151) relates how:

> During the isolation of the Great War, I came across a number of the English *Strand Magazine*. In it, among a number of pointless contributions, I read a story about a young couple who move into a furnished flat in which there is a curiously shaped table with crocodiles carved in the wood. Towards evening the flat is regularly pervaded by an unbearable and highly characteristic smell, and in the dark the tenants stumble over things and fancy they see something undefinable gliding over the stairs. In short, one is led to surmise that, owing to the presence of this table, the house is haunted by ghostly crocodiles, or that the wooden monsters come to life in the dark, or something of the sort. It was quite a naïve story, but its effect was extraordinarily uncanny.

Freud makes at least two mistakes in summarising the story, the first of which is that the young couple move into a house, not a flat, furnished only with the table ("the table goes with the house" says the agent) and the table has carvings of alligators, not crocodiles, though Moberly's story itself slips between them as crocodiles, not alligators, inhabit New Guinea which is referenced in the story. These slips may be symptomatic of something more profound and if I were Freud I would no doubt think so and would analyse them (and him) for it. I am more interested, though, in the presences and absences in Freud's account, what he reproduces correctly and what he misses out altogether. Freud is not alone in referring to alligators or crocodiles as monsters as we hear regularly in the media and as we shall see. There is a textual warrant for doing so in that the first-person narrator of the story refers to alligators as "loathly monsters." Yet this is not in relation to the carved, wooden alligators on the table, but in relation to a story within the story, a traveller's tale, about real, living alligators in a swamp in New Guinea that Freud does not consider at all as if he only read half the story by Moberly (more of that shortly).

First, the table. The first-person narrator (May, the wife of Freud's "young couple") describes the effect the carved alligators had on her:

> as the light fell on the scaly bodies they had an extraordinary look of life, and the little sinister heads with the small evil eyes almost seemed to move. I shuddered and drew away from the table. (Moberly 1917, 573, 1991, 184)

May feels quite faint for a moment and says to the agent "'there is such a queer smell in here,' [...] becoming all at once conscious of a strange and penetrating odour I had not before noticed." She begins to suggest that it might be the drains but the agent is quick to interrupt and advise that they were "'set in order before the last tenant vacated the house [...] I have the sanitary people's certificate about them'" (Moberly 1917, 573, 1991, 185). This terrace house is, however, not a working-class 'swamp' whose bad air was the supposed vector of malaria in the miasmatic theory of disease and the target of the Sanitary Movement (see Giblett 2024, chapter 4). The carved figures on the table are creatures of the swamp and smell is a vector of the uncanny in the psychoanalytic theory of the psychopathology of everyday life in modernity.

May's husband Hugh's first encounter with the table is just as uncanny as hers. He runs his fingers over the carved surface of the table and rests them on the head of one of the alligators, "a head fashioned with such skill that its loathsome naturalness made one shudder. 'Good heavens, May, the things look so lifelike I could almost have sworn one of them squirmed'" (Moberly 1917, 574, 1991, 186). Inanimate things coming to life, or seeming to do so, or imitating life, is for Freud one of the features and vectors of the uncanny.

The figuration of the alligator as 'a thing,' 'this thing' and 'the thing' persists into recent fiction as this is precisely how T. Coraghessan Boyle figures one in his 1990 novel *East is East* (Boyle 1990, 335–336). "The Thing" has been defined by Kristeva (1989, 13) as:

> the real that does not lend itself to signification, the centre of attraction and repulsion, the seat of sexuality from which the object of desire will become separated [...] the Thing is an imagined sun, bright and black at the same time.

"The Thing" occupies a zone prior to and outside signification, what Kristeva (1984) elsewhere calls "significance," which is also prior to and outside the separation of subject and subject in what Kristeva (1982) also calls abject. It is a creature of the uncanny, the centre and site of fascination and horror, the vehicle and vector of the monstrous.

Hugh and May are visited by a friend called Jack Wilding for whom the carved alligators trigger memories not related by Freud in his recounting of the story, which highlights the importance of the fact that the table is located in a house, not a flat. After dinner "on a delicious May night" the two men friends are chatting with the windows open. The spring smells of the garden waft into the room:

> when all at once the drifting sweetness from without was tainted by that same strange odour which we had noticed once or twice before ... As it drifted across the room our guest suddenly sat bolt upright in his chair, and a curious greyness overspread his naturally bronzed complexion. 'My God!' he said, what is that? And why does it smell the same – the same –' His sentence trailed off into silence, and in the intense stillness following his strange words I heard a sound which, for some reason I could not pretend to explain, gave me a feeling of cold fear. I can only describe the sound as like a far-away bellowing – not precisely the bellowing of cattle, but a more sinister, more horrible sound, pregnant with evil. 'You hear it too?' Jack Wilding questioned, under his breath ... 'And the stench is here too! Good God! If I thought I should ever have to cross that swamp again I would go mad.' (Moberly 1917, 576, 1991, 190)

The immediacy of the sense of smell takes Jack back to what was old and long familiar, and to what was long forgotten, if not repressed. Jack then recovers himself sufficiently to say:

> 'I must have had a nightmare – a waking nightmare,' he said, looking around him. 'I could have sworn that I smelt the alligator swamp in New Guinea, the place where –' he broke off short. 'I heard the loathsome brutes bellowing,' he began again; 'but, of course – or course, it was merely some association of ideas.' (Moberly 1917, 577, 1991, 190)

Hugh indicates the table and suggests that it was the trigger for Jack's association of ideas:

> Jack turned and glanced at the table, and he recoiled when he saw the grinning heads lying amongst the crusted delicacy of leaves and flowers. 'Loathsome beasts!' he said, and again his voice shook… I crossed an alligator swamp once with a friend … It was dark, the place swarmed with those unspeakable devils, their stench was everywhere. It was dark – and poor old Danson' – he paused, as if speech were almost impossible – 'they dragged him off the path of the logs in the darkness' … Somehow his words brought before me the hideous swamp, the darkness, the loathly monsters waiting for their prey, and the remembrance of just such an incident in a book I had once read flashed into my mind. (Moberly 1917, 578, 1991, 191)

The original *Strand Magazine* publication of the story has accompanying illustrations by Dudley Tennant, one of which is of this story within the story, complete with a glimpse of both the table and the log path, and a depiction of Jack as both the teller of the tale and character in it (Moberly 1917, 577).

Jack goes on to relate how "the place swarmed with those unspeakable devils'.' Swarming creatures, in biblical terms, are 'an abomination.' They are neither fish nor flesh nor fowl. They neither just swim nor walk nor fly, but do all three. These alligators are no exception. When the wooden alligators come so life they always seem to be sliding or slithering between Jack's or Hugh's feet (but not May's, as presumably she keeps her legs together like a lady). The housekeeper later describes how they go 'slithering' and 'running on their underneaths' (Moberly 1917, 574, 1991, 193). They do not walk on all fours like domesticated animals. They do not separate their grotesque lower bodily stratum from the grotesque lower earthly stratum, but are part and parcel of it. The abominable is also almost beyond words. It is inexplicable as the story concludes and as its title signals.

The normative modern relationship between monsters and reason is depicted by Francisco Goya created in his 1799 etching, 'The Sleep of

Reason Produces Monsters.'[1] It depicts a sleeping figure dreaming of monstrous owls, bats and a cat larger than a domesticated one, perhaps a feral cat or a wild feline, such as a leopard. The title and etching imply that monsters are produced in the nightmare of reason when it is sleeping. Monsters do indeed lurk in the sleep of reason and return from the monstrous repressed of reason to haunt dreams, but not because the latter produced the former. Rather, the sleep of monsters produces reason. Reason tried to put monsters to sleep. Monsters precede reason (as mythology shows). The birth of monsters precedes the birth of reason (as the Bible and the old English elegiac long poem *Beowulf* show; see Giblett 2018, chapter 2). The (putting to) sleep of monsters made reason possible. The Christian-induced sleep and repression of monsters produced the bad dream of rationalism and the Enlightenment, in which the repressed monstrous returns.[2] The uncanny both returns to the repressed monstrous in the past, and returns the repressed monstrous to presence in the present.

The Enlightenment for Sloterdijk (2009, 66) "amounted to a continuation of Christianity by rationalist and historico-philosophical means by virtue of its ideological or propagandistic means," not least against monsters and the uncanny. Ideologies for Sloterdijk (2009, 136) "in the strong sense of the word are movements that ape the form of zealous monotheism with atheistic world projects," not least with the draining of wetlands, the home of monsters and the uncanny place *par excellence*. The "para-monotheist" Enlightenment for Sloterdijk (2009, 135, 136) "not only broke away from the historically developed monotheisms; it produced a higher level monotheism in which various universal articles of faith attained dogmatic validity. These include [...] the destiny of humans to control nature." The drive for mastery over nature, rather than the desire for mutuality with it, was bred in hubris with the devastating consequences seen everywhere today.

What Byung-Chul Han (2017, 58) calls "the first Enlightenment" in which "imagination, corporeality and desire were repressed" has given way to what he calls "the second Enlightenment" of digital psychopolitics in which imagination, corporeality and desire are unrepressed, celebrated

[1] Reproduced in Castle (1995, figure 1.5, 18) with the title 'The Nightmare of Reason' without comment or analysis. See Beal (2002, 119) for a discussion of Goya's etching.

[2] As Castle (1995) shows for that movement and period. For Christianity, monsters and dragons, see Giblett (2018, chapter 2, 'Theology of Wetlands and Marsh Monsters' and chapter 3, 'Theology of Dragons and Monstrous Serpents').

and commodified by the tech giants and their algorithms with smartphones and social media. Han (2017, 58) calls for "a third Enlightenment [...] in order to shine a light on how digital enlightenment has transformed into a new kind of servitude," the major thesis of *Psychopolitics*. In the digital enlightenment smartphones for Han (2017, 12) are "the *devotional objects of the Digital* [that] work [...] like a rosary" in which "*Like* is the digital *Amen*" (his emphases). Rather than another Enlightenment that would morph into yet another barbarism as with the two previous ones as Han (2017, 58–59) points out, the metaphors of enlightenment need to be deconstructed and decolonised. For there to be enlightenment, there has to be a corresponding process of 'endarkenment.' The individual process, and cultural mo(ve)ment, of enlightenment is linked to and made possible by a concomitant process of 'endarkenment,' whether it be of the 'dark continent' of Africa or of female sexuality (or perhaps more precisely the relation to the mother[3]), or of the 'black water' of the wetland, all of which tend to get conflated and used to figure each other in the patriarchal western tradition. The second Enlightenment of digital psychopolitics is made possible by the endarkenment of 'mother earth' from which the minerals come to make the new technologies of power and which is expected to hide its wastes, but which return in the Anthropobscene.

The psychopolitics of consumerist capitalism is a not a matter of rationalist repression of the uncanny and the monstrous, but of managing them productively in daydreams placed under what Benjamin (1973, 233, 2002, 113, 2003, 261) called "the phony spell" of "the putrid magic of the commodity." What Han (2017) calls "the psychopolitics of neoliberalism" and its "new technologies of power" put the consumer asleep in daydreams that repress the monstrous Anthropobscene, but in which it lurks to return despite the best efforts of "neoliberal psychopolitics [...] not to repress" any and every thing (Han 2017, 36). Rather than neoliberalism being the new digital colonialism with its technologies of power as Han (2017, 25) claims that "discovered [and exploited] the psyche as a productive forces," consumerist capitalism and the old analogue colonialism with its communication technologies had already made and mined the discovery of the fetishism of commodities critiqued by Marx that Benjamin (1999) found alive and thriving in the glass-domed arcades of nineteenth-century Paris and that is ongoing in their neoliberal progeny of the shopping malls and online shops of today. The psychopolitics of consumerist capitalism and its

[3] For Irigaray (1993, 10) 'the relation to the mother is the 'dark continent' par excellence.'

sublime communication technologies of power from the analogue to the digital (Giblett 2008) try to hide its secret of the uncanny Anthropobscene, the wastelands, rubbish dumps and slums of cities, such as Paris and London, founded in swamps (as we will see in later chapters), the sweat shops of globalisation and the hell of global boiling. "Today" is not only what Han (2017, 11) calls "the age of digital psychopolitics," but also the age of global boiling. Psychopolitical ecology critiques and decolonises both and opens a pathway to hope through the uncanny.

5.3 Being Crocodile Prey

The return to the repressed monstrous in the past and the return of the repressed monstrous to presence in the present in the uncanny is evident in the case of the close encounter with a crocodile of the Australian eco-feminist Val Plumwood in northern Australia in 1985. She told the tale in terms that separated the good, crocodile-free waters of a beautiful wetland from the bad, crocodile-infested waters of an uncanny swamp. She also told the tale in terms of the patriarchal monster-myth despite her best efforts to keep it, and the crocodile at bay. She lived to tell the tale, but it took some time for her to be able to tell it. When humans venture into the habitat of crocodiles they can be prey and "being prey" as Val Plumwood describes it is a terrifying experience, but also instructive as it was for her. Being prey highlights the non-reciprocal nature of the relationship between prey and predator (though crocodiles are often regarded as scavengers). The relationship more specifically between eater and eaten is ultimately non-reciprocal: only one being gets to eat and the other to be eaten. Besides consuming crocodiles as meat, humans also consume their habitat by destroying wetlands in a colonising and non-reciprocal relationship.

I am not going to attempt to retell the story in all its gory details and if I attempted to sum it up by saying it was a gripping and gut-wrenching story I would be making bad puns. Plumwood's story is worth reading for itself. By reading it critically I am not belittling or demeaning her experience, which was traumatic to state the obvious, nor her individually as she was heroic to say the least. Rather, I am trying to critique the patriarchal and colonial elements that emerge in it despite her best efforts to keep them at bay. She also tells the tale in terms that separate the good, crocodile-free waters of a beautiful wetland from the bad, crocodile-infested waters of an uncanny swamp.

Her story has some explicit elements of the uncanny, such as her description of "the unfamiliar sensation of being watched" and her "whispering sense of unease" prompted, not by the sight of a crocodile, but of "a strange rock formation" (Plumwood 1996, 33–4, 1999, 78, 2000a, 57, 2000b, 130) which is a portent of what is to come. Plumwood hears whispers of unease and observes the strange rock formation so she is in a heightened state of sensory alertness. When the crocodile attacks her flimsy fibre-glass canoe "the unheard of was happening" (Plumwood 1996, 34, 1999, 78, 2000a, 131, 2000b, 57). The uncanny is evoked by hearing (or not) and smelling, by what can be heard or not, what can be smelt, but generally not by seeing. Deathly silence evokes the uncanny. In this case, being seen also evokes it. "Being prey" as she calls it, is the result of being seen. To be prey one is seen. The crocodile stalks its prey using the senses of hearing, smell and sight. Plumwood is not the subject of the gaze, but its object. She hears whispers of unease and hears (and sees) the unheard of, a crocodile attacking a canoe with her in it.

The ten-year gap between "being prey" and publishing her account and reflections on this event under this title attests perhaps to the difficulty for her of coming to terms with her experience and expressing it. Part of the difficulty was not only the trauma of the attack itself, but also the way in which her story was subjected to what she called "the cultural drive to represent it [the attack] in terms of the masculinist monster myth: the master narrative" (Plumwood 1996, 40, 1999, 85, 2000a, 139, 2000b, 59). In this myth, the crocodile is constructed as a ravening, orally sadistic monster who (or which) rapes and eats his (and it is always a male in the myth) innocent female victim. Yet Plumwood reproduces the myth herself when, for example, she earlier describes in lurid terms how she had "a blurred, incredulous vision of great toothed jaws bursting from the water" that then "seized [her] between the legs in a red-hot pincer grip" (Plumwood 1996, 35, 1999, 79, 2000a, 131, 2000b, 57). The crocodile is figured as a monstrous, orally sadistic and reptilian cousin of 'Jaws,' the monstrous shark.

Rather than only construing her experience in good ecological terms of being prey and so being a part of the food chain, Plumwood also turns in the longer version of her story to the mythology of New Guinea (and why not that of Australian Aborigines one wonders) as a way of accounting for her experience that does not merely deny or repress its mythological elements and significance. She suggests that "crocodiles are masters of water" and goes on to argue that

> the crocodile is an exploiter of the great planetary dualism of land and water. As Papua New Guinea writer Vincent Eri suggests in his novel, *The Crocodile*, the creature is a sort of magician: its technique is to steal the Other, the creature of the land, away into its world of water where it has complete mastery over it. Water is the key to the crocodile's power, and even large crocodiles rarely attack in its absence. The crocodile is then a boundary inhabitant. (Plumwood 1996, 39, 1999, 84, 2000a, 137)

In other words, the crocodile, is a wetland inhabitant, an inhabitant of the intermediary zone between dry land and deep water that crosses "the boundary" between land and water and upsets the dualism between them (as we saw in chapter 2). Yet Plumwood overlooks the fact that the wetland with its distinctive features is the crocodile's habitat. She describes the latter as "the swamp" and contrasts that with "Kakadu's wetlands" with their "dreamlike beauty" enticing her into "a joyous afternoon's idyll" at the beginning of the story. She describes how here she "glutted" herself on "the magical beauty [...] of the lily lagoons untroubled by crocodiles" (Plumwood 1996, 33, 1999, 76, 2000a, 128, 2000b, 56). She moralises the pastoral world of the wetland as good and heavenly—good enough to eat with herself as metaphorical glutton—and the swamp as bad and hellish, the place of the orally sadistic and gluttonous crocodile. For her, the crocodile lives in a swamp whereas the tourist visits a wetland; in brief, swamp is bad, wetland is good.

Naturally, she would rather eat than be eaten (and wouldn't we all?) Just as she is being watched, but prefers to watch, so she prefers to "eat" the beauty gluttonously than to be eaten greedily by the crocodile. Who can blame her? The point is, though, oral sadism rules, okay? She is a metaphorical glutton and the crocodile is figured as an orally sadistic monster. Tourist and crocodile are going about their normal, everyday business subscribing in both cases to the master narrative—despite Plumwood's best efforts to avoid it and do otherwise. The tourist watches (preferably without being watched) and consumes beauty gluttonously through his or her eyes without being consumed; the crocodile watches the tourist and sometimes consumes him or her through his or her mouth. Both are just doing their own thing in this mythology of modernity with its non-reciprocal and non-symbiotic positions of mastery of watching and eating, rather than being watched and eaten.

As the wetland for Plumwood is good and heavenly, the crocodile swamp by implication is bad and hellish, though the wetland and the swamp are one and the same body of water—they are just figured differently. The wetland is Eden before the Fall, or before the appearance of the serpent, here transformed into the crocodile. It is also the place of good magic, but crocodiles are excluded from this world of good magic as they bring trouble into paradise by enticing creatures of the land with bad magic into a watery grave, into a wet underworld, into its world—the swamp. The good magic is a sanitised white magic enticing Plumwood into the beautiful pastoral idyll of the wetland split off from, and valorised over, the black magic of the crocodile enticing her into the horrifying black water of the swamp. Magic is just magic, though, without the moralisation. Plumwood (curiously for an ecofeminist) reproduces not only the patriarchal, western moralisation of the wetlandscape, but also its dualisms and spatial metaphysics and poetics of land and water, good and bad, white and black, heaven and hell, above and below

5.4 Religion, Swamps, and Thoreau

The uncanny is religion without (genitive and material) the boundaries of mere reason. The uncanny does not require reason and has absented itself from it. It is contrary to Kant's "religion within the boundaries of mere reason" that he differentiated in the 1790s from historical, institutionalised religions (Kant, 1998). This is not to imply that the uncanny is irrational; it is pre-rational and post-rational, before and after reason historically and chronologically, and ex-rational, outside and after reason conceptually and experientially. In the 1940s Georges Bataille invoked Kant for the title of the second part of *Theory of Religion* by entitling it 'Religion within the Limits of Reason' without acknowledging, referencing or discussing Kant (Bataille 1992, 7, 63–104). For Bataille as an erstwhile associate of the French surrealists who were aficionados of that which is outside the limits of reason, such as dreams, this seems bizarre.

In the conclusion to *Theory of Religion* Bataille repeated as a throwaway provocation the title of Goya's etching without attribution, comment or elaboration (Bataille 1992, 113). In the body of this book, Bataille (1992, 57) defined religion in Proustian terms as "the search for lost intimacy." One wonders, intimacy with what? With God, time, place, nature, animals, plants, people, bodies, gods, or goddesses? Or all of them? Intimacy, as Bataille implies also in Proustian terms, does not necessarily have a fixed object, nor a stable subject, but is, in Kristeva's (1982) terms, abject. Religion is not the search for a lost object, including a lost time, which

Proust only found fleetingly when a madeleine and his involuntary memory transported him back to the past.

Religion is the search for lost intimacy *of* and *between* time, place, nature, animals, plants, people, bodies, gods, goddesses and God. The uncanny is the vehicle and vector that enacts this intimacy —whether it is searched for or not. The uncanny can come searching for the dreamer—awake or asleep. The uncanny for Sloterdijk (2023, 230) "exists and dwells among us. Concrete religions are styles of the uncanny." By contrast, artists for him are "the ecologists of uncanny things" (Sloterdijk 2017, 190). Artists (including writers) are high and low priests, not stylists, of the uncanny living among us in the earthly household. These writers include Henry David Thoreau, the greatest thinker of new America, the patron saint of swamps, and a pioneer of conservation counter-theology and psychopolitical ecology.[4]

A century before Bataille, Henry David Thoreau (1958, 52, 2013, 89) defined religion more affirmatively than Bataille when he wrote in a letter that "our religion is where our love is." God is not love (contra the New Testament Epistle of 1 John 4: 8 in which "God is love"), but religion is love, or more precisely where our love is. When one finds love, one finds religion. When one loses (a) love, as Thoreau did when his beloved older brother died from lockjaw (tetanus) and his sweetheart rejected him on her father's orders, one can become lost, as Thoreau did for a while. The uncanny is the search for lost love, or lost love searching one out, often horrifically, as it did for Thoreau who developed sympathetic symptoms of his brother's lockjaw and who never married. The uncanny is not the search for a lost object of love (as that has gone and is lost), but the search for (lost) love. Thoreau (1982, 613) found love (and so religion) in a swamp:

[4] There are some remarkable comparisons and contrasts to be drawn between Thoreau and William James, including what Sloterdijk (2020, 225) calls James's 'religious-philosophical project of a liberal empiricism [... ditto Thoreau]; his New England Protestantism of formative education, his scientifically minded Harvard discipline [ditto Thoreau, who also attended Harvard], his non-obtrusively impassioned individualism [ditto Thoreau, though he was obtrusive and delighted in 'obtruding' on his readers] and his euphoric prose [ditto Thoreau who is considered one of the finest American writers of euphoric prose ...] [James's] path was one of empathic and excited meta-religiousness.' Ditto Thoreau. James seemed to be following in Thoreau's footsteps. For further discussion of Thoreau as 'the patron saint of swamps,' see Giblett (1996, xi, 229–239, 2020, chapter 4, 2021, chapter 9, 2025) and a pioneer of conservation counter-theology, see Giblett (2018).

> When I would recreate myself, I seek the darkest wood, the thickest and most interminable, and to the citizen, most dismal swamp. I enter a swamp as a sacred place – a *sanctum sanctorum* [holy of holies]. There is the strength, the marrow of Nature,

of the body of the earth and his own body. Sacrality, sanctuary, psycho-political ecology and conservation counter-theology go hand-in-hand in a four-handed couple.

Thoreau was writing in the 1850s when what David Miller (1989, 3) calls "unconscious mental processes" were "newly awakened" in a fascination with, and horror of, swamps. It was also a period of political turmoil with conflict over slavery and the lead up to the American Civil War in the following decade. It was also a period of environmental destruction of habitats, such as woods and swamps. Thoreau was also writing in this period to awaken citizens to unconscious natural and political processes, especially those associated with woods and swamps. Like Benjamin, Thoreau was at the forefront of the invention of psychopolitical ecology with diagnoses of the symptoms of pscyhogeopathology and the prescription of psychosymbiosis as the prevention or cure in theory and practice, such as the swamp cure of immersion. Thoreau wrote in his journal that "far from being poisoned in the strong water of the swamp, it is a sort of baptism for which I had waited" (Thoreau 1962, *IX*, 376–377).

Religion is what binds people and place together in love, both people loving places and places loving people, as Rebecca Solnit has argued. Sense of place for Solnit (2013, 30) leads to, or is more precisely and affectionately, "love of place," not only "our love for places," but also "how [...] places love us back" and appreciating "what they give us." Place for Solnit (2014, 1) is "the intersection of many changing forces passing through [...] a fixed location." In a place "ecology, democracy, culture, storytelling, urban life, individual life histories and collective endeavour coexist geographically, spatially." Solnit (2014, 3) celebrates "the pleasures and possibilities each place holds [...] with the beauties of light, space, and solidarity." Thoreau is an exemplary writer of place in all these regards (see Giblett 2025, 2026).

Thoreau's swamp religion, however, is not where love is to be found for some good citizens. In mainstream western culture swamps and other wetlands, such as marshes, mires, morasses, bogs, lagoons, sloughs, shallow lakes and estuaries, etc., have been seen traditionally, or at least in patriarchal times, as places of darkness, disease and death, horror and the uncanny, melancholy and the monstrous—in short, as black waters. They have often been regarded as home to some sort of horrifying marsh monster or swamp serpent lurking in their murky waters. Up until the 1890s it was thought that the miasmatic vapours that rose from stagnant pools caused malaria (which means literally 'bad air'). The perception persisted from ancient times that miasma also cause melancholia or depression. Wetlands have been filled or drained in order not only to prevent malaria and melancholia, but also because they pose an obstacle to agricultural and urban development.[5] Swamps are where hate and horror is for some, and where melancholy and depression were not for Thoreau. The uncanny is the name for this ambivalent love/hate religious relationship with monstrous places, such as swamps or slums, and/or with monstrous beings, such as alligators or crocodiles.

The uncanny is a religious category, concept and experience, or more precisely it is a religious concept or experience for a culture and time in which God is dead and so akin to the sublime. As the sublime is a secular theology, so is the uncanny; as the sublime stands in for God as a secular theology (perhaps an oxymoron), so the uncanny stands in for the monstrous divine as a secular demonology; as the sublime is a secular God-substitute, so is the uncanny a secular Satan-substitute; as Satan is a fallen angel, so the uncanny is the fallen angelic for the world in which God is dead. The sublime and the uncanny can also enact the search for lost intimacy and love in and through technology, or more precisely through sublime communication technologies, such as photography and cinema.

Religion for Thoreau is where love is before God stands there and so melancholy cannot stand there. Melancholy for Thoreau does not stand in the swamp, or in nature more generally. On October 31, 1857 Thoreau wrote in his *Journal* (1962, *X*, 150) that "if you are afflicted with melancholy [...], go to the swamp." Thoreau flies in the face of the convention that swamps are melancholic places (see Giblett 1996, chapter 7). In *Walden* he wrote that "there can be no very black melancholy to him who lives in the midst of Nature and has his senses still" (Thoreau 1982, 382). Living in the middle of nature (as Thoreau did for some of his life and time) is the prophylactic for melancholy in a world in which God, or one's beloved brother, or love for a sweetheart, is dead, but love lives. There can be no black melancholy to him or her standing in the middle of a swamp

[5] Each of these aspects of wetlands receives a systematic, often chapter-by-chapter, treatment in Giblett (1996).

and who has his or her senses still to absorb local sounds, smells and sights. Thoreau (1962, *I*, 141) wrote in his *Journal* that "would it not be a luxury to stand up to one's chin in some retired swamp for a whole summer's day." When one stands in a swamp, melancholy is not there, though the uncanny (the (un)homely) may be.

For these reasons Thoreau found hope in a swamp in the Anthropobscene. He announced in his late essay "Walking" that

> hope and the future for me are not in lawns and cultivated fields, not in towns and cities, but in the impervious and quaking swamps [...] I derive more of my subsistence from the swamps which surround my native town than from the cultivated gardens in the village [...] A town is saved, not more by the righteous men in it than by the woods and swamps that surround it. (Thoreau 1982, 611–613)

The alligator for Thoreau (1982, 543) "comes out of the mud with quakings of the earth." Thoreau goes into the quaking zones of the earth, such as swamps, where he finds love and hope. Thoreau (1962 *XIV*, 301–302) wrote in his journal about "the open swamp [...] where the surface quakes for a rod around" that "there is no wilder and richer sight than is afforded from such a point of view of the edge of a [...] swamp." Hope, not melancholy, is in the quaking and uncanny swamp. Thoreau did not despair, a topic of the following chapter, as is melancholy.

References

Bachofen, J. J. 1967. *Myth, Religion and Mother Right: Selected Writings.* Translated by Ralph Manheim. Princeton, NJ: Princeton University Press.

Bataille, Georges. 1992. *Theory of Religion.* Translated by Robert Hurley. New York: Zone. First published in French in 1973.

Beal, Timothy. 2002. *Religion and Its Monsters.* New York: Routledge.

Benjamin, Walter. 1973. *Illuminations.* Translated by Harry Zohn. London: Fontana.

Benjamin, Walter. 1999. *The Arcades Project.* Translated by Howard Eiland. Cambridge, MA: Belknap Press of Harvard University Press.

Benjamin, Walter. 2002. *Selected Writings: Volume 3, 1935–1938*, ed. Howard Eiland and Michael Jennings. Cambridge, MA: Belknap Press of Harvard University Press.

Benjamin, Walter. 2003. *Selected Writings: Volume 4, 1938–1940*, ed. Howard Eiland and Michael Jennings. Cambridge, MA: Belknap Press of Harvard University Press.

Boyle, T. Coraghessan. 1990. *East Is East.* New York: Penguin.

Castle, Terry. 1995. *The Female Thermometer: Eighteenth-Century Culture and the Invention of the Uncanny.* New York: Oxford University Press.

Douglas, J. D., ed. 1970. *New Bible Dictionary.* London: Inter-Varsity Press.

Freud, Sigmund. 1955. The 'Uncanny'. In *The Standard Edition of the Complete Psychological Works of Sigmund Freud, Volume XVII (1917–1919): An Infantile Neurosis and Other Works*, 217–256. Translated by Alix and James Strachey. London: Hogarth. First published in German in 1919.

Freud, Sigmund. 1957. The Antithetical Meaning of Primal Words. In *The Standard Edition of the Complete Psychological Works of Sigmund Freud, Volume XI (1910): Five Lectures on Psycho-Analysis, Leonardo da Vinci, and Other Works*, 154–161. Translated by James Strachey. London: Hogarth. First published in German in 1910.

Freud, Sigmund. 1975. Findings, Ideas, Problems. In *The Standard Edition of the Complete Psychological Works of Sigmund Freud, Volume XXIII (1937–1939): Moses and Monotheism and Other Works*, 299–300. Translated by James Strachey. London: Hogarth. First published in German in 1938.

Freud, Sigmund. 1985. The 'Uncanny'. In *Art and Literature, Penguin Freud Library, Volume 14*, 335–376. Translated by Alix and James Strachey. Harmondsworth: Penguin. First published in German in 1919.

Freud, Sigmund. 2003. *The Uncanny.* Translated by David Mclintock. London: Penguin. First published in German in 1919.

Giblett, Rod. 1996. *Postmodern Wetlands: Culture, History, Ecology.* Edinburgh: Edinburgh University Press.

Giblett, Rod. 2008. *Sublime Communication Technologies.* Basingstoke: Palgrave Macmillan.

Giblett, Rod. 2018. *Environmental Humanities and Theologies: Ecoculture, Literature and the Bible.* London: Routledge.

Giblett, Rod. 2019. *Psychoanalytic Ecology: The Talking Cure for Environmental Illness and Health.* London: Routledge.

Giblett, Rod. 2020. *New Lives of the Saints: Twelve Environmental Apostles.* Lanham, MD: Hamilton Books.

Giblett, Rod. 2021. *Wetlands and Western Cultures: Denigration to Conservation.* Lanham, MD: Lexington Books.

Giblett, Rod. 2024. *Wetland Cultures: Ancient, Traditional, Contemporary.* New York: Palgrave Macmillan.

Giblett, Rod. 2025. *Waking to Nature with Thoreau and Benjamin: Or, Psychopolitical Ecology.* New York: Bloomsbury.

Giblett, Rod. 2026. *The Politics of Nature and Environmental Writing: Political Ecocriticism*. Palgrave Macmillan.
Han, Byung-Chul. 2017. *Psychopolitics: Neoliberalism and New Technologies of Power*. Translated by Erik Butler. London and New York: Verso.
Irigaray, Luce. 1993. Body Against Body: In Relation to the Mother. In *Sexes and Genealogies*, 9–21. Translated by Gillian C. Gill. New York: Columbia University Press.
Kant, Immanuel. 1998. *Religion within the Boundaries of Mere Reason and Other Writings*, Edited and translated by Allen Wood and George di Giovanni. Cambridge: Cambridge University Press. First published in German in 1793.
Kelly, Lynne. 2006. *Crocodile: Evolution's Greatest Survivor*. Crows Nest, NSW: Allen and Unwin.
Kristeva, Julia. 1982. *Powers of Horror: An Essay on Abjection*. Translated by Leon Roudiez. New York: Columbia University Press. First published in French in 1980.
Kristeva, J. 1984. *Revolution in Poetic Language*. Translated by Margaret Waller. New York: Columbia University Press. First published in French in 1974.
Kristeva, J. 1989. *Black Sun: Depression and Melancholia*. Translated by Leon. Roudiez. New York: Columbia University Press. First published in French in 1987.
Miller, David. 1989. *Dark Eden: The Swamp in Nineteenth-Century American Culture*. Cambridge: Cambridge University Press.
Moberly, L. G. 1917. Inexplicable. *The Strand Magazine* 54 (324): 572–581.
Moberly, L. G. 1991. Inexplicable. In *Strange Tales from The Strand*, ed. Jack Adrian, 183–195. Oxford: Oxford University Press. First published 1917.
Parker, Philip. 2015. *The Northmen's Fury: A History of the Viking World*. London: Vintage.
Plumwood, V. 1996. Being Prey. *Terra Nova* 1 (3): 32–44.
Plumwood, V. 1999. Being Prey. In *The New Earth Reader: The Best of Terra Nova*, ed. D. Rothenberg and M. Ulvaus, 76–91. Cambridge, MA: MIT Press.
Plumwood, V. 2000a. Being Prey. In *The Ultimate Journey: Inspiring Stories of Living and Dying*, ed. J. O'Reilly, S. O'Reilly, and R. Sterling, 128–146. San Francisco, CA: Travelers' Tales.
Plumwood, V. 2000b. Being Prey. *UTNE Reader*, July–August, pp. 56–61.
Quammen, David. 2004. *Monster of God: The Man-Eating Predator in the Jungles of History and the Mind*. London: Hutchinson.
Randolph, Jeanne. 2001a. Looking Back at Cyborgs. In *The Uncanny: Experiments in Cyborg Culture*, ed. Bruce Grenville, 182–186. Vancouver: Arsenal Pulp Press.
Randolph, Jeanne. 2001b. Transgressed Boundaries: Potent Fusions and Dangerous Possibilities. In *The Uncanny: Experiments in Cyborg Culture*, ed. Bruce Grenville, 95–99. Vancouver: Arsenal Pulp Press.

Royle, Nick. 2003. *The Uncanny*. New York: Routledge.

Sloterdijk, Peter. 2009. *God's Zeal: The Battle of the Three Monotheisms*. Translated by Wieland Hoban. Cambridge, UK: Polity. First published in German in 2007.

Sloterdijk, Peter. 2011. *Spheres, Volume 1: Bubbles: Microspherology*. Translated by Wieland Hoban. Los Angeles, CA: Semiotext(e). First published in German in 1998.

Sloterdijk, Peter. 2017. *The Aesthetic Imperative: Writings on Art*, ed. Peter Weibel. Translated by Karen Margolis. Cambridge, UK: Polity Press. First published in German in 2014.

Sloterdijk, Peter. 2020. *After God*. Translated by Ian Alexander Moore. Cambridge, UK: Polity Press. First published in German in 2017.

Sloterdijk, Peter. 2023. *Making the Heavens Speak: Religion as Poetry*. Translated by Robert Hughes. Cambridge, UK: Polity. First published in German in 2020.

Solnit, Rebecca. 2013. *The Faraway Nearby*. New York: Viking.

Solnit, Rebecca. 2014. *The Encyclopedia of Trouble and Spaciousness*. San Antonio, TX: Trinity University Press.

Thoreau, Henry David. 1958. *The Correspondence of Henry David Thoreau*, ed. Walter Harding and Carl Bode. Washington Square, NY: New York University Press.

Thoreau, Henry David. 1962. *The Journal of Henry D. Thoreau, Volumes I–XIV*, ed. Bradford Torrey and Francis Allen. New York: Dover.

Thoreau, Henry David. 1982. *The Portable Thoreau*, ed. Carl Bode. New York: Viking.

Thoreau, Henry David. 2013. *The Correspondence of Henry D. Thoreau: Volume 1: 1834–1848*, ed. Robert Hudspeth. Princeton, NJ: Princeton University Press.

Trudinger, Peter. 2001. Friend or Foe? Earth, Sea and *Chaoskampf* in the Psalms. In *The Earth Story in the Psalms and the Prophets*, ed. Norman Habel, 29–41. Sheffield: Sheffield Academic Press.

Vollmar, Fritz. 1972. Preface. In *Crocodiles: Their Natural History, Folklore and Conservation*, ed. Charles Guggisberg, ix–x. Mount Eliza, VIC: Wren.

CHAPTER 6

Mourning, Melancholy, and Despair

Abstract Besides the uncanny, Freud also took up melancholy and mourning in an essay on both of them in which he argues that mourning arises from the loss of a significant other and melancholy from the loss of the self or other valued object. Melancholy is associated with death, or at least with the (v)alley of the shadow of death (the precursor to death) and with mourning (the aftermath of death). Melancholy is the immanent counterpart to the transcendental sublime and the spiritual counterpart to the psychological uncanny. As the sublime and uncanny are a secular theology for a world in which God is dead, so melancholy is a secular spirituality in the lower psychopathological register. Melancholy is the neurosis symptomatic of capitalist society driven by greed for, and indulging in gluttony of, the earth's resources in the Anthropobscene, rather than exercising gratitude for her generosity in the desire for mutuality. This chapter goes on to discuss Walter Benjamin's work on the German mourning play, E. M. Cioran's figuring of despair as mud, Wilkie Collins's evocation of the black water of despair, and Charles Dickens's portrayals of "mudfog" and "marshy commons" in England and "the grim domains of giant despair" in an American swamp.

Wetlands were once thought to be bad for the body. According to the miasmatic theory of disease, malaria—literally 'bad air'—could be caught by breathing miasma or bad air. Wetlands could also be bad for the mind.

R. Giblett, *Living with Nature in the Anthropobscene*,
https://doi.org/10.1007/978-3-032-12822-5_6

Indeed, they can plunge the mind into melancholia, and even into madness and despair. Melancholy is the immanent counterpart to the transcendental sublime and the spiritual counterpart to the psychological uncanny. As the uncanny is a secular theology for a world in which God is dead, so melancholy is a secular spirituality in the lower psychopathological register of what Julia Kristeva (1982) calls the abject, the mediating category between subject and object that makes both possible.

Melancholy is associated with death, or at least with the (v)alley of the shadow of death (the precursor to death) and with mourning (the aftermath of death). Melancholia and mourning are connected in Sigmund Freud's essay on the topic, in Walter Benjamin's work on the German mourning play, in E. M. Cioran's discussions of despair and melancholy, and in Charles Dickens's depiction of "the grim domains of Giant Despair" as the present chapter goes on to discuss.

In a typical assessment made more than a century ago the later-to-be President of the United States Theodore Roosevelt (cited by Brooks 2014, 111) referred to "the melancholy marshes" in the "lonely lands" of "the wilderness." More specifically, Hubert Davis (1962, 21) referred to "the stench of a rotting carcass of some depraved hunter lost in the depths of its [the Great Dismal Swamp's] melancholy jungles."[1] The corpse beneath the swamp is not only a reminder of death, but also a vector of melancholy. This assessment of the marsh as melancholic can also be found in the work of those masters of dejection, the Romantic poets. Percy Bysshe Shelley (1975, 51–52), for example, referred to "a wide and melancholy waste/ Of putrid marshes" when evoking "the Spirit of Solitude." Previously Samuel Pepys (1972, 311) in the late seventeenth century travelled "over most sad Fenns (all the way observing the sad life that the people of that place [...] do live)." A century ago the map of "the world" of those 1920s classics of Anglophone children's literature, A. A. Milne's *Winnie-the-Pooh* and *The House at Pooh Corner*, with its Christopher Robin of the clean and proper body, has that arch manic-depressive donkey Eeyore living in "Eeyore's Gloomy Place" which is "rather boggy and sad" (Milne 1958, frontispiece map). Other examples in adolescent literature are the Swamps of Sadness in Michael Ende's *The Neverending Story* and the Dead Marshes in J. R. R. Tolkien's *The Lord of the Rings* (see Giblett 2018, chapter 2).

[1] For the Great Dismal Swamp on the border between Virginia and North Carolina and its writers, beginning with William Byrd in the eighteenth century, see Giblett (1996).

Despondency is another name for melancholy that is the immanent counterpart to the transcendental sublime and the spiritual counterpart to the psychological uncanny. As the uncanny is a secular theology for a world in which God is dead, so melancholy is a secular spirituality in the lower psychopathological register of the abject, Kristeva's (1982) term for the mediating category between subject and object that makes both possible. In 1933 Gershom Scholem (Benjamin and Scholem 1992, 81) included a poem in a letter to Walter Benjamin the last line of which concludes that "where God once stood now stands: Melancholy." Melancholy is the abject spirituality of a world in which God is dead; hope in a living swamp is the abject spirituality for a world in which religion is love.

Melancholy is also, as Terry Eagleton (1986, 41) argues, "the appropriate neurosis for a profit-based society." Capitalist society is driven by greed for, and indulges in gluttony of, the earth's resources. Gluttony is one of the seven deadly sins in the Christian tradition. In a collection of "antidotes to a culture of greed," Sloterdijk (2014, 11) prescribes thinking "a society of generosity," but acknowledges that "a politics of generosity remains a dream." A generosity-based society and environ-mental health go hand-in-hand. A profit-based society and environ-mental illness remain an interlocked nightmare. Rather than gratitude for the generosity of the great mother earth, greed and gluttony are directed towards her in the psychogeopathologies of mining and anal sadism, and pastoralism and oral sadism (Giblett 2011, chapters 9 and 10; 2019, chapters 5 and 6).

6.1 Mourning, Melancholy, and Freud

Just as the aesthetic and philosophical sublime lifts one up to the heady heights of intellection and theory close to the divine, so the psychological and spiritual melancholy of the uncanny depresses one down into the grotesque lower bodily and earthly strata of slime close to the demonic beneath what Gérard de Nerval calls "the black sun of melancholia" (see Kristeva 1989; see also Giblett 1996, chapters 2 and 7; 2013, chapter 14). Melancholy is associated with death, or at least with the (v)alley of the shadow of death (the precursor to death) and with mourning (the aftermath of death). Melancholia and mourning are connected for Freud. In his 1917 essay, 'Mourning and Melancholia,' Freud argued that mourning is "commonly the reaction to the loss of a beloved person," whereas melancholia "may be a reaction to the loss of a beloved object" (Freud 2005, 203 and 205). The loved objects which are lost for the melancholic is specifically the breasts of the mother and the water which is breast milk or the water of the wetland, the first water which nourished life (on earth), the breasts of mother earth.

More broadly, the loved and lost object can be 'nature' generally as Freud (2005, 195–200) identified and discussed in his essay 'On Transience' published in 1916, the year before 'Mourning and Melancholia.' For Freud (2005, 199), mourning and melancholia were particularly acute during the destruction of World War I that "robbed the world of its beauties" and "destroyed [...] the beauty of its landscapes" (as we will see in the following chapter). Freud (2005, 198) found some consolation in the cycle of the seasons in which "the beauty of nature [...] returns the following year after the ravages of winter" and in the fact that "that return may be seen as eternal in terms of the length of our lives." The melancholic ego, however, is immersed in the moment and does not have the time-scale of the eternal. Freud (2005, 197) describes a young poet who demonstrates the affliction of, and fixation on, transience that is the topic of the essay and the impetus for writing it. The young poet values transient objects because of their "scarcity over time" and so experiences the "painful world-weariness" of melancholia. By contrast, Freud values enduring objects, such as antiquities, because of their generosity over time and so experiences the pleasurable world-liveliness of eternality.

Instead of investing desire in the objects of love, such as the breasts and/or the mother and gaining some return of pleasure on that investment (the economic metaphor is appropriate), the melancholic ego "wants to incorporate this object into itself, and in accordance with the oral or cannibalistic phase of libidinal development in which it is, it wants to do so by devouring it," or by simply "eating it" in the less orally sadistic terms of the more recent translation (Freud 1984, 258, 2005, 210). In environmental terms, the melancholic wants to incorporate the nourishing qualities of the living waters of the wetland breasts into himself (the melancholic is usually male) by devouring them through drainage or filling, and even by creating artificial ones. The object of investment was initially an object of love which was later lost.

Freud goes on to distinguish between mourning and melancholia by arguing that "in mourning it is the world which has becomes poor and empty; in melancholia it is the ego itself" (Freud 1984, 254, 2005, 205–206). In mourning the world is experienced as loss, whereas in melancholia the ego is experienced as lost. Freud outlined the process whereby melancholia establishes an "identification of the ego with the abandoned object" (Freud 1984, 258, 2005, 209): "the shadow of the object fell upon the ego [...] as though it were an object, the forsaken [or "abandoned"] object. In this way the loss of the object was transferred into a loss of ego" (Freud 1984, 258, 2005, 209). In melancholy, as Mikkel Borch-Jacobson (1989, 183) concludes in his discussion, "the ego becomes the object," or even more succinctly, "'the lost object [is] me'" as Nicolas Abraham and Maria Torok (1984) put it, as a way of denying or disavowing the loss of the loved object in narcissistic inversion.[2]

6.2 Melancholy, Despair, and Cioran

Melancholy differs from despair. In melancholy the ego is an object of ambivalent love/hate for the subject, whereas in despair the ego becomes an object of enmity and hatred for the subject. Despair in these terms is put no more eloquently and succinctly than in Shakespeare's *Richard III* (2: 2, 36–37) when the widowed Queen (whose husband, King Edward IV, was murdered) says, "I'll join with black despair against my soul. And to myself become an enemy." Cioran uses these two lines from *Richard III* as the epigraph to the 'Directions for Decomposition,' the first part of *A Short History of Decay* (Cioran 2010, 3). In despair, the lost object of the ego is the enemy, an object of hate for the subject, whereas in melancholy the ego and the lost object are one and the same in ambivalent self-love/hate. Melancholy and despair also differ in colour-coding in mainstream western culture. For Cioran (1995, 43), the doyen of dark despair and the master muse of modern melancholy, "blue is a soothing color for melancholy," whereas black is the distressing color of despair.[3]

For Cioran (1995, 15) despair, "more than any other feeling, establishes a correspondence between our being and the environment. In fact,

[2] Renee Aron Lertzman (2013, 124–126) in her discussion of mourning and melancholia in Freud's essay 'On Transience' (and not in his essay on these topics) considers "the lost or damaged object," such as "a body of water," as outside and separate from the subject, and not as (part of) the subject. Lertzman (2015, 117) also invokes "Freudian melancholia" in her book-length study of environmental melancholia, but does not discuss Freud's essay on mourning and melancholia which seems like a lost opportunity to engage with their psychoanalytic dimensions and to relate them to environmental issues and concerns in psychopolitical ecology.

[3] The association between blue, such as the blue flower of Romanticism, and melancholy is discussed further in Giblett (2021, chapter 8). Black is discussed further in Giblett (2013, chapter 14).

despair requires a corresponding environment to such an extent that if needs be, it creates it." Despair festers in the feral quaking zone of the dead black waters of industrial wastelands. Confronted with the horror of the filthy, smelly mud on the Western Front in World War I produced by the terror of artillery bombardment dispensed by the monstrous machines of industrialised warfare, the self/land is made to hate itself and war is fought against the self/land (as we will see in the following chapter). The wastelands of the cities reduced to rubble and ruins in World War II produced by the terror of firebombing dispensed also by the monstrous machines of industrialised warfare followed suit and fitted the bill too (we will see in Chap. 8). More recently, in what Antoine Picon (2000) calls "the anxious landscape" of "regions of ruin and rust" in the steel belts of post-industrial wastelands (depicted on the cover of *Living with Nature in the Anthropobscene*), industrial technology and infrastructure decay and decompose in pools of stagnant, and possibly toxic, water. All three are landscapes of the Anthropobscene.

Like despair, melancholy also establishes a correspondence between human beings and 'the environment' (natural or cultural), requires a corresponding environment and, if needs be, creates it (in wetland tropes) with 'the slough of despond' and with what Cioran (2010, 110) figures as "languid mud" in a section 'On Melancholy' of *A Short History of Decay*. Melancholy for Cioran (2010, 110; his emphases) is "the *dream state of egoism*: no longer any object outside oneself, no reasons for hate or love, but that same fall into a languid mud, that same circling of the damned without a hell." By contrast, in the blackness of despair oneself is the object of hate inside oneself (as the widowed Queen puts it in *Richard III*). Languid mud for Cioran is the poetic dreamscape of melancholy and the dream space necessary for it.

Cioran also associates the swampy city of Venice with muddy melancholy. For Cioran (1995, 92), "Venice is not a historic reality but a function of melancholy, a town of tears caught between doubts and dreams." It is also caught between dread and death, as Vera Brittain demonstrates on both counts (as we will see in the following chapter). Melancholy requires a corresponding environment and, if needs be, creates it, not only in languid mud, but also in wetland cities, such as Venice, caught between the dreams of draining and the doubts of drowning. Melancholy produces the draining dream of the eternal and serene city (as with Venice), as well as the drowning nightmare of the decaying and decomposing city that was built in wetlands. In the case of Venice, 'the *alta aqua*' of floods caused increasingly by rising sea levels in the age of global heating brings the historic reality of the muddy, melancholic city back to the surface from the swamp from where it came in the return of the repressed.[4]

Cioran figures melancholy not only as languid mud functioning in the swampy city of Venice, but also in miasmatic and malignant tropes in a secular hell. Melancholy for Cioran (2010, 110) "requires a debauch of space, an infinite landscape in order to spread out its sullen and vaporous grace, its shapeless evil." Cioran is secularising and seems to be paraphrasing Dante's description of the fifth circle of hell with the souls of the sullen stuck in slime in "a dreary swampland, vaporous and malignant" (as Dante puts it in canto 7 of his *Inferno* and as we saw in Chap. 3). The secular modern melancholic is damned for the time being to, and within, a swampy hell figured in denigratory mainstream terms; they have a circle of hell and an environment in it. Secularised modern melancholy requires and, if needs be, creates the infinite landscape of the figuratively debauched and morally and aesthetically ambiguous swamp of Cioran's dream-scape. Without this dream-scape, without this surface or screen to project its dream state, or nightmare, of egoism onto, melancholic death in life and ambivalent self-love/hate would not have a place to breed the self-hatred of black despair that can lead to the end of life in self-death. Cioran is not only "the existentialist of refusal" as Sloterdijk (2013, 76) calls him, but also a psychogeophobe of mud and a diagnostician of the Anthropobscene.[5]

The languid mud of melancholy mixes the elements of earth and water; melancholy is muddy so it flows slowly; melancholy is stagnant. Melancholy for Cioran (2013, 85; his emphasis) is not only "languid mud," but also "*stagnant* blood." Whether it is languid brown mud, billowing black bile, stagnating red blood or slimy white phlegm, melancholy, like water in and through wetlands, flows slowly. In the theory of the humors, melancholy is associated with bile, phlegm with the phlegmatic, and blood with the

[4] For further discussion of Venice as a wetland city, see Giblett (2016, chapter 5).

[5] Despite bidding "farewell to philosophy" when "it became impossible to discover in Kant and in all the philosophers any human weakness, any authentic accent of melancholy," Cioran (2010, 49, 54–55) still entertained enlightenment philosophical metaphors, such as when he claims we "habitually we sink into a nocturnal mud, into a darkness quite as mediocre as the light […] Life […] is a monotonous loss of light, an insipid dissolution in the darkness." There would be no process of enlightenment or movement of 'the Enlightenment' without a corresponding and concomitant process and movement of 'endarkenment,' in Cioran's case, of the muddy and living black waters in the heart of darkness. Cioran still turned towards philosophy in turning away from it. He was Janus-faced, part of his fascination.

sanguine. They all mediate solid and liquid. They can also be in the process of transformation from liquid to solid as they dry up, such as slimy phlegm and coagulating blood, or in transition from solid to liquid (or vice versa), such as muddy melancholy.

Cioran's languid mud of melancholy is a secularised slough of despond with strong theological roots that he acknowledges explicitly in a later section of *A Short History of Decay* devoted to the 'Genesis of Melancholy.' Melancholy has its beginnings, or "genesis," for him in the fall from paradise as told in the story in the biblical book of Genesis. Cioran (2010, 143) suggests that melancholy "harks back to the root of our ruin" in the fall from paradise and he concludes that melancholy is "the poetry of original sin." Although melancholy is not strictly one of the seven deadly or cardinal sins in the Christian tradition, it is linked to the deadly sin of pride or egotistical self-love.[6] Cioran (2010, 110) figures it as the "strangest flower of self-love." It is the counterpart to the blue flower of impossible romanticist love. It is also linked to the deadly sins of sloth or laziness, and greed or gluttony (making up three out of the seven). It can lead to despair (the one unforgiveable sin), self-hate (anger or wrath is another of the seven deadly sins) and suicide (the ultimate act of egotistical self-love and a sin, partly because the perpetrator is not able to ask for forgiveness afterwards). Whereas wrath is anger directed against another person, melancholy is anger or hate (and ambivalently pride or love) directed towards oneself, and greed and gluttony directed towards the earth. All these sins flow from the original sin of disobeying God's interdiction against eating the fruit of the tree of knowledge of good and evil. No sin*s* without the singular original sin of knowing sin by consuming the fruit of God's creation. Melancholy is associated with so many other sins that it is a multi-faceted mega-sin rather than a cardinal sin as such.

Cioran secularises the fifth circle of Dante's poetic hell in which melancholics, with "black bile billowing" inside them (following the theory of the humors as we saw in Chap. 2; see also Pensky 2001, 22–23 and 252, n.39), are "bitter in the blackened mud" of a swamp (as Dante puts it in keeping with the denigration of wetlands). They have fallen into the slime of a swamp where the souls of the sullen are stuck (as we saw in Chap. 3). The sullen have committed the deadly sin of anger against another and

[6] Max Pensky (2001, 31) suggests that "as a cardinal sin, melancholia proceeds as a direct consequence of the Fall." The former assumption is mistaken theology I suggest, while the latter diagnosis concurs with Cioran.

they embody wrath. The melancholic commits the deadly sins of pride and self-anger, and embodies self-love/hate. They are stuck in the same circle as the sullen of the medieval slime of hell for Dante, secularised as the modern hell of slime by Jean-Paul Sartre and others (as we saw in Chap. 3 too).

Hatred resides in the figurative wetland of what Leonard Cohen, the musical master of modern melancholy, called "the swamps of hate" of racism and sexism. They are linked to the placism of the hatred of swamps (misaquaterrism) that describes wetlands in denigratory terms with their sullen miasmas, poisonous vapours, sucking slimy mud, entangling vegetation, dirty, disease-ridden water, and evil and orally sadistic monsters (imaginary or real, such as alligators and crocodiles) lurking in the depths of their dead black waters waiting to devour greedily the unwary who venture into their watery/earthy domain of slime (as we have seen in the previous chapter).

Love, by contrast, lives and thrives in the native quaking zone of living black waters with their monstrous creatures and their fertile and fetid water that invite wading into the slimy swamp when it is safe to do so, a sacred space and place in which to self-baptise and evade the heights and depths of despair, cultivate gratitude for their generosity and faith in life—even in death—and nurture hope (as Henry David Thoreau found, and as we saw in Chap. 5). Melancholic ambivalent self-love/hate mediates these two wetlandscapes as Cioran suggests in his figuring of melancholy as languid mud and as morally and aesthetically ambiguous.

The infinite wetlandscape of ambivalent melancholy arises from the obverse of the infinite spatiality of the sublime (as we saw in Chap. 3). Without the sublime, melancholy would not be possible. Without the secular theology of the sublime, the fall into the secular demonology of melancholy would not be possible. Without sublimation into the abstract and surmountainous heights of despair, philosophy and theory, the fall into the depths of slime in a secular hell and the rise into the black sun of miasmatic melancholy would not be possible. The sublime is sometimes accompanied with melancholy as Kant said (as we also saw in Chap. 3). Both sublime and melancholy entail upward displacement, or artificial transcendence. Both accompany each other on upwardly parallel paths, or trajectories as it were. Both differ from the infinite temporality, or eternality as in Freud's discussion, of the wetland.

6.3 The Black Waters of Wetlands

The wetland figured as 'black water' with all its incipient racist associations would be a convenient short hand label to encapsulate more than two millennia of patriarchal western vilification and destruction of wetlands. An exemplary instance of this figuring is found in Wilkie Collins's *The Woman in White* of 1859–60 set mainly in Blackwater Park with its Blackwater Lake. Marian Halcombe takes a walk in the Park and finds herself:

> standing suddenly on the margin of a vast open space, and looking down at [...] Blackwater lake [...] The lake itself had evidently once flowed to the spot on which I stood, and had been gradually wasted and dried up to less than a third of its former size. I saw its still, stagnant waters, a quarter of a mile away from me in the hollow, separated into pools and ponds by twining reeds and rushes, and little knolls of earth. On the farther bank from me the trees rose thickly again, and shut out the view, and cast their black shadows on the sluggish, shallow water. As I walked down to the lake, I saw that the ground on its farther side was damp and marshy, overgrown with rank grass and dismal willows. The water, which was clear enough on the open sandy side, where the sun shone, looked black and poisonous opposite to me, where it lay deeper under the shade of the spongy banks, and the rank overhanging thickets and tangled trees. The frogs were croaking, and the rats were slipping in and out of the shadowy water, like live shadows themselves, as I got nearer to the marshy side of the lake. I saw here, lying half in and half out of the water, the rotten wreck of an old overturned boat, with a sickly spot of sunlight glimmering through a gap in the trees on its dry surface, and a snake basking in the midst of the spot, fantastically coiled and treacherously still. Far and near the view suggested the same dreary impressions of solitude and decay, and the glorious brightness of the summer sky overhead seemed only to deepen and harden the gloom and barrenness of the wilderness on which it shone. I turned and retraced my steps to the high heathy ground. (Collins 1974, 227–228)

This passage reproduces much of the mainstream western iconography and phenomenology of the wetland. The marshy side of the lake is figured as a kind of Eden-after-the-Fall in which a satanic serpent lurks. Yet Dante and Milton had gone even further than Collins in satanising the swamp as part of their theologising of the landscape as we have seen in earlier chapters: Dante by figuring one circle of hell as a slimy Stygian marsh, and Milton by troping Satan as a monstrous swamp serpent who is generated out of the slime of hell.

What is probably more remarkable, though, about this passage from *The Woman in White* than its reproduction of traditional Christian iconography, and its moralisation of the marsh, its *marécage moralisé*, is the trajectory it traces through space from the high heathy, even healthy, ground descending spatially and morally to the low poisonous, decidedly unhealthy wet-land, and the juxtaposition it sets up between the sublime sky and the slimy swamp; between the openness of the former and the claustrophobia of the latter; the life and light of the former and the death, darkness, decay and disease of the latter; between the glories of the former and the waste-land of the latter.

The narrator/character of Marian Halcombe is positioned as first person in relation to the landscape, or more precisely what could be called the wetlandscape, or at least its visual space, entirely via the sense of sight. The repetitious insistence on, and of, "I saw" places her in relation to various "views," or lack of them, the landscape reduced to occasions for taking, or thwarting, "views." She is not represented as hearing, touching, smelling or tasting the wetlandscape but is placed in a position of mastery from which to look down on the wetland in order to know it and so in order to master it, even though it is waste land. In fact, the "I" is arguably constituted as bourgeois, indeed petit bourgeois, individual, albeit masculinised, subject insofar as the waste/wet land is postulated as alien other.

The figuring of the wetland as wasteland is exacerbated for Marian one windy and cloudy morning when "the rapid alternations of the shadow and sunlight over the waste of the lake made the view look doubly wild, weird, and gloomy" (Collins 1974, 253). The surface of the water reflects and 'heightens' the depths of gloom of the sky on a dark and overcast day, whereas on a bright and glorious day it is juxtaposed with it. The surface of water and the depths of the sky are set up with the petit-bourgeois individual consciousness situated between, and mediating, them. Philosophers may endlessly debate whether the crash of a tree in a forest occurs if there is no individual consciousness to perceive it, but for there to be a reflection in water there has to be a visually perceptive individual consciousness to receive it.

Indeed, the individual consciousness is constituted in modern European landscape painting not only as the point of intersection of the lines of perspective, but also as the terminus of the lines of reflection. The surface of water, Henri Lefebvre (1991, 86) has argued, "symbolises the surface of consciousness and the material (concrete) processes of decipherment which brings what is obscure forth into the light," or into the open.

Consciousness deciphers the marks on the surface of the water and brings the obscure into transparency, dark into light. The reduction of the water to surface for decipherment also denies, if not represses, the depths of the water. Water, Luce Irigaray (1985, 289) argues, "serves as a reflecting screen and not as a reminder of the depths of the mother; it sends back the image of the sun, of men, of things," as depicted on the cover of *Living with Nature in the Anthropobscene*. The individual consciousness reflecting on the surface of things is a patriarchal consciousness which represses the depths of the mother. It necessarily finds the wetland boringly repetitive unless it is broken up by pools of reflective water. In Rider Haggard's *She* there were:

> wide stretches of lonely, death-breeding swamp, unbroken and unrelieved so far as the eye could reach, except here and there by ponds of black and peaty water that mirror-like, flashed up the red rays of the setting sun. (Rider Haggard 1991, 65–66)

The patriarchal consciousness is here able to situate itself in relation to the datum point of the sun via the occasional reflective pool of water, albeit black in which what Julia Kristeva (1989) calls "the black sun of melancholia" is reflected. Without these markers it/he would be in danger of being lost in the depths of the swamp monster/mother.

Yet when the individual consciousness is set up in this reflexive, or more precisely self-reflexive, relationship between water and sky, it is constituted as narcissistic, and as self-seduced, paranoiac of absorption by the mother's body, or by bodies of water. The mirror of water, Jean Baudrillard (1990, 69) argues, is "not a surface of reflection, but of absorption [...] It is always a matter of self-seduction [...] All seduction in this sense is narcissistic." Reflection on the surface of things in the reflections of the surface of the water is, in other words, narcissistic and self-seductive.

Given the horrors of the depths of the swamp to the patriarchal consciousness it is only 'natural' or inevitable that the wetland wasteland is quite useless from a capitalist agricultural point of view, though it may be picturesque from an aesthetic point of view and yet at the same time the perfect setting for a murder, and so associated with illegal death, three facts which are associated in Sir Percival's mind in *The Woman in White* when he relates how:

> 'Some people call that picturesque,' said Sir Percival, pointing over the wide prospect with his half-finished walking-stick. 'I call it a blot on a gentleman's property. In my great-grandfather's time the lake flowed to this place. Look at it now. It is not four feet deep anywhere, and it is all puddles and pools. I wish I could afford to drain it, and plant it all over. My bailiff (a superstitious idiot) says he is quite sure the lake has a curse on it, like the Dead Sea. What do you think, Fosco? It looks just the place for a murder, doesn't it?' (Collins 1974, 253)

For Sir Percival the wetland is a blot, a black mark, on a gentleman's property, on its clean and proper surface, on his character, on its title deeds and in its ledgers, and not even a debit in red. Sir Percival refers to this blot in the twin indexical gestures of the demonstrative "that" (with no following substantive as if it were indescribable) and pointing with his stick as if he were unable to describe it adequately, or even get it into words.

Indeed, the wetland constitutes the unutterable in the maintream western tradition, hence the recourse that is invariably made, and Marian Halcombe as a narrator of *The Woman in White* is no exception, to the lexicon of what could be called standard swampspeak with its vocabulary of such well-worn words as "dismal" (Collins 1974, 279), "dreary" (228, 253 and 280), "desolate" (280), "gloomy" (253) and so on. Wetlands, perhaps more than any other site, bear out Jacques Derrida's (1981, 160) contention that "we find ourselves unable to represent 'place' itself except by metaphors." Wetlands have almost invariably been represented in the patriarchal western tradition in metaphors of despair and despondency in a overworking of the lower echelons of the empathetic fallacy and of the nether regions of the psycho(geo)pathological register.

Equally we find ourselves equally unable to represent 'states' or abstractions, or at least our stance towards them, except by metaphors of 'place' or sites. Often our stance towards an ideological or political position is figured in terms of our stance towards a place. Wetlands have been used metaphorically to convey our repudiation or vilification of the ideologically incorrect in an overworking of the lower reaches of the psychopolitical register. A favourite expression in the vocabulary of critical theory is the word 'mire,' either as noun or verb.[7] Capitalism, colonialism or patriarchy, for example, are figured as a mire. To take an ideological misstep

[7] See, for example, Eagleton's (1991, 91 and 97) references to "the mire of historical relativism" and to "a bourgeois class sunk in the mire of immediacy."

and to be mired in one of them, or what's worse in a combination of two of them, or even all three (horror of horrors), is to find oneself sunk irretrievably off the straight and narrow of the ideologically correct. Yet the pejorative use of the mire metaphor is implicated ('mired'!) in the capitalist, colonialist, and patriarchal repudiation and vilification of wetlands. Such theorists would never dream of using a racist or sexist metaphor to figure the ideologically incorrect, but persist in the use of 'placist' metaphors of wetlands and so are what could be called 'misaquaterrists,' haters of wetlands.

Besides recourse to the merely indexical, Sir Percival has recourse to the imperatives of capitalist agriculture which dictate that the wetland is useless land as it stands and that it should be drained and transformed into productive land. For Sir Percival water under four feet deep, a rough rule of thumb for defining the upper limit of water in a wetland, is by definition useless, as are puddles and pools, almost too shallow for a wetland except perhaps for seasonal or temporary wetlands which could even be dry for much of the year. The wetland could be viewed as picturesque, but never as beautiful, and certainly not ever as producing the sublime, the pinnacle (the mountain metaphor is appropriate) of aesthetic pleasure, albeit bordering on pain in the case of the sublime, the upper echelons of the empathetic fallacy and the higher reaches of the psychopathological register. At the other end of the scale, wetlands constitute, by and large, the obverse, or secret as Zoë Sofoulis puts it, of the sublime which she construes as the slimy and encapsulates in the parenthetical portmanteau s(ub)lime (as we saw in a previous chapter).

6.4 The Grim Domains of Giant Despair and Melancholy

Despair and melancholy have been figured in slightly differing wetland tropes with melancholy as the slough of despond and despair as a wetwasteland. In the early modern vision of hell granted to Saint Teresa of Avila (1957, 233) in a moment of despair "the ground appeared to be covered with a filthy wet mud, which smelt abominably and contained many wicked reptiles." Filthy mud, bad smells, and evil monsters are associated with despair and hell by Saint Teresa.

Charles Dickens follows suit and does likewise in the nineteenth century with "the grim domains of Giant Despair" in chapter 23 of *Martin*

Chuzzlewit. Without the advantage of having available Cioran's outline of the differences in theory between despair and melancholy, Dickens conflates them and highlights the difficulty of keeping them apart in practice, though both for him reside in, and are associated with, a wetland. The final leg of the journey of Martin Chuzzlewit to Eden at the confluence of the Ohio and Mississippi Rivers is through a scene made up entirely of:

> sky, wood and water, all the livelong day; and heat that blistered everything it touched. On they toiled through the great solitudes, where the trees upon the banks grew thick and close; and floated in the stream; and held up shrivelled arms from out [*sic*] the river's depths; and slide down the margin of the land, half growing, half decaying in the miry water. On through the weary day and melancholy night: beneath the burning sun, and in the mist and vapour of the evening: on, until return appeared impossible, and restoration to their home a miserable dream.

This is a journey into a place where the elements are mixed, a place which gives birth to death or to a land where the trees are half-growing and half-dying. This mixture of the elements and life and death gives birth to "the dull depression of the scene." The journey is a secularised epic descent into the underworld on "old Charon's boat, conveying melancholy shades to judgment." Judgment of whom and for what is not specified, but the post-epic journey through the swamp as modern secular hell is part of the test which Martin must complete to achieve manhood on his journey to return home.

From this point the situation can only deteriorate as these are only the approaches to the underworld; this is not Hell itself but only the journey across the River Styx:

> as they proceeded further on their track, and came more and more towards their journey's end, the monotonous desolation of the scene increased to that degree, that for any redeeming feature it presented to their eyes, they might have entered, in the body, on the grim domains of Giant Despair. A flat morass, bestrewn with fallen timber; a marsh on which the good growth of the earth seemed to have been wrecked and cast away, that from its decomposing ashes vile and ugly things might rise; where the very trees took the aspect of huge weeds, begotten of the slime from which they sprung, seeking whom they might infect, came forth at night, in misty shapes, and creeping out upon the water, hunted like spectres until day; where even the

> blessed sun, shining down on festering elements of corruption and disease, became a horror; this was the realm of Hope through which they moved.

Giant Despair is a character straight out of Bunyan's *Pilgrim's Progress* who owns "Doubting-Castle" (Bunyan 2008, 116). Dickens transforms despair into the characteristic of capital 'D' Despair of gigantic proportions and transfers the giant's domain from a castle into a dismal swamp and slough of despond.

Although for Dickens Eden is ironically not a paradise but a festering sink in which the elements are improperly mixed and from this improper mixing spring hideous and monstrous deformities and mutants, by a further countervailing irony the wetland chaos could be seen to be God's first work, not so much Eden, the pastoral paradise, but the pristine pre-Edenic, pre-pastoral and pre-paradisaical wetland (Giblett 2024, chapter 3). Dickens' Eden, however, is far from that watery chaos. It is both a "hideous swamp" "choked with slime" and a "dreary situation" that produces "dread" and "the apprehension of death." Like the approaches to the swamp-town in which the elements of heat, water, air and earth are mixed, so too in Eden itself they are mixed where "a fetid vapour, hot and sickening as the breath of an oven, rose up from the earth, and hung on everything around."

At Eden Martin could not leave his footprints on "the marshy ground" for as soon as his foot sank into it, "a black ooze started forth to blot them out." Martin is also unable to leave much of a mark on the land, or more precisely wetland, as it was "mere forest" whose:

> trees had grown so thick and close that they shouldered one another out of their places, and the weakest, forced into shapes of strange distortion, languished like cripples. The best were stunted, form the pressure and want of room; and high about the stems of all, grew long rank grass, dank weeds, and frowsy underwood: not devisable into their separate kinds, but tangled all together in a heap; a jungle deep and dark, with neither earth nor water at its roots, but putrid matter, formed of the pulpy offal of the two, and of their own corruption.

This wetland is a nightmare for the taxonomising botanist who would seek to separate out its specimens of flower, fruit and leaf into genus and species. This wetland is not just the excrement of earth and water, but the corrupt, decomposing excrement of earth and water. More generally, wetlands are a taxonomic anomaly in a classificatory order predicated on a hard and fast distinction between land and water, time and space.

Such is the strength of the association between the wetland and the horrific in the modern European cultural unconscious that when Dickens would portray the evils and pollutions of the Victorian industrial city he figures it as an unhealthy marsh. In one of his little known works, *The Mudfog Papers*, the title and eponymous town encapsulates the taxonomic anomaly of the wetland: mud as mixing earth and water, fog as a mixture of air and water. Water is thus a kind of promiscuous substance that gets around, even sleeps around, too much with earth and air, especially in the industrial town, that sink of iniquity:

> water is a perverse sort of element at the best of times, and in Mudfog it is particularly so. In winter, it comes oozing down the streets and tumbling over the fields, – nay, rushes into the very cellars and kitchens of the houses, with a lavish prodigality that might well be dispensed with; but in the hot summer weather it <u>will</u> dry up, and turn green: and although green is a very good colour in its way, especially in grass, still it certainly is not becoming to water; and it cannot be denied that the beauty of Mudfog is rather impaired, even by this trifling circumstance. Mudfog is a healthy place – very healthy; damp, perhaps, but none the worse for that. It's quite a mistake to suppose that damp is unwholesome: plants thrive best in damp situations, and why shouldn't men [*sic*]? [...] So, admitting Mudfog to be damp, we distinctly state that it is salubrious. The town of Mudfog is extremely picturesque. (Dickens 1987, 1)

The town is picturesque despite, or because of, in Dickens's ironical terms, its green water in summer. The irony upon irony is that it is a "vulgar error," certainly in the terms of twentieth-century scientific and medical discourse, though certainly a persistent one in popular culture, that damp causes disease.

"The health-preserving air of Mudfog" is precisely not health-preserving, not because of its dampness, but because of the pollutants being spewed out into its water and air by its factories to form the mudfog of its name. Like the fog figured as a yellow cat in T. S. Eliot's 'The Waste Land,' the fog of Mudfog has a life of its own as it rises from the modern urban wasteland figured as wetland:

> it had risen slowly and surely from the green and stagnant water with the first light of morning, until it reached a little above the lamp-post tops; and there it had stopped, with a sleepy, sluggish obstinacy, which bade defiance to the sun, who had got up very blood-shot about the eyes, as if he had been at a drinking party over night, and was doing his day's work with the worst possible grace. The thick damp mist hung over the town like a huge gauze curtain. All was dim and dismal. (Dickens 1987, 11)

In other words, all was like being in a wetland as it is conventionally figured and as Dickens figured it in *Great Expectations*: "the dark flat wilderness [...] was the marshes" which were, in turn, "the dismal wilderness." Richard Lehan (1998, 44) notes how in Dickens's later novels the uncanny is evoked and enacted in the liminal space of the marsh:

> Between the country and the city is a strange, eerie, primitive world of the marshes – a world of water and mire with houses sinking into the mud, a world of sluice gates and mills. The narrative flash points in Dickens's fiction occur where water and land meet, or where the country and city intersect, or where the past and present converge.

And where the former and latter are figured in terms of each other. Similarly Robert Alter (2005, 53) refers to Dickens's "aqueous urban landscape of iron, rust and rot," as depicted again on the cover of *Living with Nature in the Anthropobscene*. Or more precisely, the waste wetlandscape of the Anthropobscene that begins with the industrial city, its polluted air, land and water, and typified in its slimy slums (as we will see in Chap. 9).

Yet rather than a later development in Dickens's novels (as Lehan argues), he was fascinated with wet wastelandscapes over the course of his long career as in *The Mudfog Papers* (Dickens 1987). This fascination began with his earliest published writings, such as in his description of Walworth in 'The Black Veil' first published in 1836:

> The back part of Walworth, at its greatest distance from town, is a straggling miserable place enough, even in these days; but, five-and-thirty years ago, the greater portion of was little better than a dreary waste [...] Very many of the houses which have since sprung up on all sides [...] were of the rudest and most miserable description. The appearance of the place [...] was not calculated to raise the spirits [...] or to dispel any feeling of anxiety or depression [...] [The] way lay across a marshy common, through irregular lanes, with

> here and there a ruinous and dismantled cottage fast falling to piece with decay and neglect. A stunted tree, or pool of stagnant water, roused into sluggish action by the heavy rain of the preceding night, skirted the path occasionally [...] so much of the prospect as could be faintly traced through the cold damp mist which hung heavily over it, presented a lonely and dreary appearance perfectly in keeping with the objects we have described. (Dickens 1994, 363–364)

This prospect was not the pleasing prospect so beloved of rural landscape writers and painters, but a displeasing prospect on the marshy margins between city and country in which "mud and mire" (Dickens 1994, 364) predominate. The pool of stagnant water was the stock-in-trade of nineteenth-century writing about the city as it stood for a source of disease in the prevailing miasmatic theory of disease and sanitary practices of inspection.

6.5 Mourning, Melancholy, and Benjamin

Along Freudian lines, Ilit Ferber (2013, 20) reiterates in a recent discussion of Walter Benjamin's (1977, 2019) work on the German mourning play (*Trauerspiel*) that in melancholy "the lost object continues to exist, now as part of the dejected subject." The lost wetlands destroyed by cities continue to exist, now as part of the dejected subject of the citizen (the denizen of the city) who maintains "a relationship with an absent lost object" (Ferber 2013, 43), just as the subject maintains a relationship with absent lost wetlands by being oblivious to them or by retelling their stories, by remarking their presence in the past on maps and their absence in the present on maps too. The aim thereby is that "the relationship between subject and object is overturned" (Ferber 2013, 46). Subject and object become what Kristeva (1982, 1–2) calls (the) abject, or what Friederich Nietzsche called the body (see Giblett 2008, 3–5). The subject and object return to the pre-subject and pre-object phase in which both were abject. The abject precedes the subject and the object. It is the third party that made both possible. The dejected subject in no longer dejected, but abjected and is no longer subject, but abject.

This de-subjectification, de-objectification, desublimation and abjectification contrasts with what Ferber (2013, 59) describes as "the destructive nature of the melancholic's response to loss" in which "the melancholic devours his lost love-object in order to retain it; in demonstrating his endless loyalty, he destroys it. Therein lies the paradox: the only way to retain the object is to destroy it." The melancholic devours his lost love-object of wetlands by using monstrous drainers and dredgers in order to retain them (literally behind retaining walls as in some cities such as Venice) and retrain them and their aberrant water (as in the city of Venice in canals; see Giblett 2016, chapter 5), but in doing so kills the wetland as a living being, as a habitat for other living beings. He kills the thing he loves, rather than loving the thing he kills as indigenous cultures do (see Giblett 2011, 215). For Ferber (2013, 73) "the melancholic's lost object is either dead (in the case of human loss) or absent (in the case of a more abstract loss)." In the case of the destruction and loss of a wetland, the melancholic's lost object is both dead (a case of non-human loss) *and* absent (a case of a concrete loss).

Furthermore, for Ferber (2013, 73), "in both cases the pathology lies in the subject's inability to recognize the loss and the insistence of maintaining the dead object as 'half-alive' within the melancholic consciousness, thus rendering the boundary between life and death indefinite and thus indistinct." In the case of the destruction and loss of a wetland, the psychogeopathology lies not only within what Ferber outlines, but also within not recognising the wetland as a place of both life and death in which life and death are mixed, but wherein the boundary between them is definite and thus distinct. The melancholic's half-alive lost object for Ferber (2013, 102) is "what is not yet dead but no longer alive." This object is buried alive in a tomb when, Ferber (2013, 102) goes on to relate, "the melancholic carves out an internal tomb for his lost object, engendering an internal topography in which the living ego and the dead object coexist," just as in Venice the city goes on living above its entombed dead lagoons of once living wombs of wetlands (see Giblett 2016, chapter 5).

In the concrete cases of the subject's relation to non-human living objects, such as animals and wetlands, the living ego and the dead object coexist by the living ego continuing to live by killing the object (see Giblett 2008, 119–120; Giblett and Tolonen 2012, 41 and 45). They are entombed as dead object, but they were once life-giving wombs. Death, as Ferber (2013, 104) concludes, "does not mark the end of life [...]: the two states exist concomitantly," as they do in the wetland's ecology. Moreover, for Ferber (2013, 115) "the dead are never completely dead, and the past can never be hermetically closed." The lost wetlands are never completely dead, and the past is still an open book, as it is in Venice (see

Giblett 2016, chapter 5). Venice is the wetland city *par excellence*, not only in terms of its past and its history as a city founded in a swamp, but also in the present and its current culture as a rich source of metaphor and of psychological travail in which these metaphors are symptomatic of its mournful and melancholic psychogeopathology.

Instead of seeing the object (breasts, wetlands) as lost, the melancholic ego sees itself as lost in a massive act of narcissistic disavowal and egotism. As a result, the ego desires itself, or in the terms of Deleuze and Guattari (1977, 26) "it is the subject [...] that is missing in desire, or desire that lacks a fixed subject; there is no fixed subject unless there is repression." Repression is to subjectivity as drainage is to wetlands; repression (and drainage) fix the flows of embodied subjects (and wetlands). Repression is constitutive precisely of *melancholic* subjectivity; subjectivity is melancholic and mournful. The subject desires itself as a product of a melancholic loss of the loved object of the mother and mother earth, of the mother's breasts and mother earth's breasts, the living waters of wetlands.

As wetlands are increasingly lost from the world and are lost as an object of love which nourishes life, both mourning and melancholia are experienced and exercised in relation to them. The world should be in mourning for the loss of its wetlands which gave it life and nourished it, but instead of being in mourning and regarding the world as losing its wetlands becoming empty or "bereft" of them, in Gerard Manley Hopkins (1953, 51) apt word, it experiences this loss as a melancholic loss of its own ego, its own selfhood and sense of identity. Selfhood and identity is constituted in relation to something outside itself against which it stands in contradistinction. When that object is lost, as when wetlands are lost, that loss is experienced in melancholia as a loss of the subject. The world is becoming empty of wetlands; the world is losing its wetlands. In patriarchy (the rule of the fathers) and filiarchy (the rule of the sons) wetlands are the bad breast of the world in which, in Melanie Klein's terms, the infantile sons try to hide their excrements, but are laid bare in the Anthropobscene. Wetlands in psychoecological terms are the womb and good breast of the world, the organs of receptivity and bounty as Klein puts it—they are living waters and living waters matter.

References

Abraham, Nicolas, and Maria Torok. 1984. A Poetics of Psychoanalysis: 'The Lost Object – Me'. *SubStance* 43:3–18.

Alter, Robert. 2005. *Imagined Cities: Urban Experience and the Language of the Novel*. New Haven, CT: Yale University Press.

Baudrillard, Jean. 1990. *Seduction*. Translated by Brian Singer. London: Macmillan.

Benjamin, Walter. 1977. *The Origin of German Tragic Drama*. Translated by John Osborne. London: New Left Books.

Benjamin, Walter. 2019. *Origin of the German Trauerspiel*. Translated by Howard Eiland. Cambridge, MA: Harvard University Press.

Benjamin, Walter, and Gershom Scholem. 1992. *The Correspondence of Walter Benjamin and Gershom Scholem, 1932–1940*. Edited by Gershom Scholem and Translated by Gary Smith and Andre Lefevere. Cambridge, MA: Harvard University Press.

Borch-Jacobson, Mikkel. 1989. *The Freudian Subject*. Translated by Catherine Porter. London: Macmillan.

Brooks, Paul. 2014. *Speaking for Nature: The Literary Naturalists from Transcendentalism to the Birth of the American Environmental Movement*. Mineola, NY: Dover.

Bunyan, John. 2008. In *The Pilgrim's Progress*, ed. Roger Pooley. London: Penguin. First published in 1678.

Cioran, E. M. 1995. *Tears and Saints*. Translated by Ilinca Zarifopol-Johnston. Chicago, IL: University of Chicago Press. First published in Romanian in 1937.

Cioran, E. M. 2010. *A Short History of Decay*. Translated by Richard Howard. London: Penguin. First published in French in 1949.

Cioran, E. M. 2013. *The New Gods*. Translated by Richard Howard. Chicago, IL: University of Chicago Press. First published in French in 1969.

Collins, Wilkie. 1974. *The Woman in White*. Harmondsworth: Penguin. First published 1859–60.

Davis, Hubert. 1962. *The Great Dismal Swamp: Its History, Folklore and Science*. Rev. ed. Self-published.

Deleuze, Gilles, and Felix Guattari. 1977. *Anti-Oedipus: Capitalism and Schizophrenia*. Translated by Roger Hurley, Mark Seem, and Helen Lane. New York: Viking Press.

Derrida, Jacques. 1981. *Dissemination*. Translated by Barbara Johnson. London: Athlone.

Dickens, Charles. 1987. *The Mudfog Papers*. Gloucester: Alan Sutton. First published 1880.

Dickens, Charles. 1994. The Black Veil. In *The Dent Uniform Edition of Dickens' Journalism: Sketches by Boz and Other Early Papers 1833–39*, ed. Michael Slater, 359–368. London: J. M. Dent. First published 1836.
Eagleton, Terry. 1986. *William Shakespeare*. Oxford: Basil Blackwell.
Eagleton, Terry. 1991. *Ideology: An Introduction*. London: Verso.
Ferber, Ilit. 2013. *Philosophy and Melancholy: Benjamin's Early Reflections on Theater and Language*. Stanford, CA: Stanford University Press.
Freud, Sigmund. 1984. *On Metapsychology: The Theory of Psychoanalysis*, Pelican Freud Library 11. Harmondsworth: Penguin.
Freud, Sigmund. 2005. *On Murder, Mourning and Melancholia*. Translated by Shaun Whiteside. London: Penguin.
Giblett, Rod. 1996. *Postmodern Wetlands: Culture, History, Ecology*. Edinburgh: Edinburgh University Press.
Giblett, Rod. 2008. *The Body of Nature and Culture*. Basingstoke: Palgrave Macmillan.
Giblett, Rod. 2011. *People and Places of Nature and Culture*. Bristol: Intellect Books.
Giblett, Rod. 2013. *Black Swan Lake: Life of a Wetland*. Bristol: Intellect Books.
Giblett, Rod. 2016. *Cities and Wetlands: The Return of the Repressed in Nature and Culture*. London: Bloomsbury.
Giblett, Rod. 2018. *Environmental Humanities and Theologies: Ecoculture, Literature and the Bible*. London: Routledge.
Giblett, Rod. 2019. *Environmental Humanities and the Uncanny: Ecoculture, Literature and Religion*. London: Routledge.
Giblett, Rod. 2021. *Wetlands and Western Cultures: Denigration to Conservation*. Lanham, MD: Lexington Books.
Giblett, Rod. 2024. *Wetland Cultures: Ancient, Traditional, Contemporary*. New York: Palgrave Macmillan.
Giblett, Rod, and Juha Tolonen. 2012. *Photography and Landscape*. Bristol: Intellect Books.
Haggard, H. Rider. 1991. *She*. Oxford: Oxford University Press. First published 1887.
Hopkins, Gerard Manley. 1953. *Poems and Prose*. Harmondsworth: Penguin.
Irigaray, Luce. 1985. *Speculum of the Other Woman*. Translated by Gillian C. Gill. Ithaca, NY: Cornell University Press.
Kristeva, Julia. 1982. *Powers of Horror: An Essay on Abjection*. Translated by Leon Roudiez. New York: Columbia University Press. First published in French in 1980.
Kristeva, Julia. 1989. *Black Sun: Depression and Melancholia*. Translated by Leon Roudiez. New York: Columbia University Press. First published in French in 1989.

Lefebvre, Henri. 1991. *The Production of Space*. Translated by Donald Nicholson-Smith. Oxford: Blackwell.

Lehan, Richard. 1998. *The City in Literature: An Intellectual and Cultural History*. Berkeley, CA: University of California Press.

Lertzman, Renee. 2013. The Myth of Apathy: Psychoanalytic Explorations of Environmental Subjectivity. In *Engaging with Climate Change: Psychoanalytic and Interdisciplinary Perspectives*, ed. Sally Weintrobe, 117–133. London: Routledge.

Lertzman, Renee. 2015. *Environmental Melancholia: Psychoanalytic Dimensions of Engagement*. London: Routledge.

Milne, A. A. 1958. *The World of Pooh Containing Winnie-the-Pooh and the House at Pooh Corner*. London: Methuen.

Pensky, Max. 2001. *Melancholy Dialectics: Walter Benjamin and the Play of Mourning*. Amherst, MA: University of Massachusetts Press.

Pepys, Samuel. 1972. *The Diary, Volume IV, 1663*, ed. Robert Latham and William Matthews. London: G. Bell.

Picon, Antione. 2000. Anxious Landscapes: From the Ruin to the Rust. *Grey Room* 1, 65-83. Accessed September 29, 2006. http://www.gsd.harvard.edu/people/faculty/picon/texts.html.

Saint Teresa of Avila. 1957. *The Life of Saint Teresa of Avila by Herself*. Translated by J. M. Cohen. London: Penguin. First published in Spanish in 1588.

Shelley, Percy Bysshe. 1975. *The Complete Poetical Works, Volume II, 1814–1817*, ed. Neville Rogers. Oxford: Clarendon Press.

Sloterdijk, Peter. 2013. *You Must Change your Life: On Anthropotechnics*. Translated by Wieland Hoban. Cambridge, UK: Polity. First published in German in 2009.

Sloterdijk, Peter. 2014. What Does a Human have that he can Give Away? In *Giving and Taking: Antidotes to a Culture of Greed*, ed. Joke Brouwer and Sjoerd van Tuinen, 9–25. Translated by Sebastian Olma. Rotterdam: V2_Publishing.

CHAPTER 7

Earth and Water in the Slime of World War I Trench Warfare

Abstract The mixing of the elements of earth and water took place in the slime and mud hell of World War I trench warfare. First-hand accounts of those who were there and wrote about it focus on these qualities. The transformation in both world wars of the natural and cultural environments in the country and the city and shifts in the conduct of industrialised world warfare indicate a change in the conduct of war at the fundamental level of the four elements of earth, air, fire and water. In the case of the WWI, with its heavy artillery and trench warfare, earth and water were mixed to produce what commentators and caption writers called "swamps," while war photographers documented and dramatized them. The transformation of the air took also place in both wars with aerial weapons. Air became an area of conflict in the third dimension of height (rather than a front in the two dimensions of length and breadth on the land). This shift took place with the development and deployment of chemical weapons during the first year of WWI. These transformations had a dramatic and devastating physical and mental impact on those who experienced them directly and indirectly.

R. Giblett, *Living with Nature in the Anthropobscene*,
https://doi.org/10.1007/978-3-032-12822-5_7

7.1 Warfare

The direct effect of the two World Wars on people has been often remarked upon. Autobiographies, histories, memoirs and novels of World Wars I and II document physical and mental trauma. They were, and still are, an important part of the publishing landscape with a steady stream of publications including: continued publication or republication of classics in English, including film tie-ins (Blunden 2000; Brittain 1978; Graves 1929; Manning 1999); new translations of French and German WWI classics (Barbusse 2004; Remarque 1996); a newly discovered first-hand account (Lynch 2006) and a recently rediscovered 'novel' (Céline 2024); a new anthology of WWI poetry (Walter 2006); a recent book of Hubert Wilkins's photographs, including those of the Western Front (Maddern 2021, 20–37); and secondary texts, such as monumental new histories of WWI (Carlyon 2006) and surveys of primary texts (for WWI see Sherry 2005). Similarly with WWII, a spate of books (Friedrich 2006; Grayling 2006; Knell 2003; Lowe2007; Taylor 2004) reflected on allied bombing of German cities on or around the 60th anniversary (discussed in the following chapter).

"The First and Second World War" for Sloterdijk (1987, 202) were "the first and second modernization of mass death." Both wars not only reshaped the nature of war and the bodies and lives of the men and women who fought these wars and who were affected and killed by them; they also reshaped the spaces, places and the time in which they were fought. WWI for David Lloyd (1998, 95) "transformed the landscape of the places over which it is fought." The same could be said for WWII. Trenches for Keith Helphand (2006, 23) "epitomized the landscape of World War I," whereas bombed cities epitomized the landscape of WWII.

The transformation of the air took also place in both wars with aerial weapons. Air became an area of conflict in the third dimension of height (rather than a front in the two dimensions of length and breadth on the land). This shift took place with the development and deployment of chemical weapons during the first year of WWI (Sloterdijk 2009, 10, 2016b, 86). For Sloterdijk (2009, 100, 2016b, 174) "the atmosphere itself became a theatre of war. More: the air becomes a weapon and a battlefield of a peculiar kind." With gas warfare, air for Sloterdijk (2009, 109, 2016b, 181) "lost its innocence," though it had already lost it with miasma, malaria and anthropogenic air pollution. The development of what Sloterdijk (2009, 9, 20, 2016b, 85, 95) calls "this battle conducted in the atmospheric environment" gave rise to what he also calls "atmoterrorism," In an interview Sloterdijk (2016a, 162) coined the term "atmospherocide" as "the typical modern form of a war of extermination." Gas warfare in WWI against soldiers led to the development of what Sloterdijk (2009, 47, 2016b, 119) calls "genocidal gas extermination," or "exterminism," during WWII against civilians in gas chambers. Air becoming an area of conflict in war during WWI also led to the development and deployment of aerial bombing and atmoterrorism in WWII against civilians in cities where for Sloterdijk (2009, 55, 2016b, 127) "the primary objective was the annihilation of civilian life" (as we will see in the following chapter).

The development of twentieth-century warfare for Sloterdijk (2009, 13, 14, 16, 2016b, 89, 90) marks "the introduction of the environment into the battle between adversaries [...] targeting no longer the body, but the enemy's environment [...] with assaults on the environmental conditions of the enemy's life." These assaults, in turn, assaulted the body of the enemy, both the soldier in WWI, and the soldier and civilian in WWII. Twentieth-century 'Total War' against the soldier and the civilian was also "an environmental war" (Sloterdijk 2009, 100, 2016b, 174) against their environment and the more-than-human environment. For Sloterdijk (2009, 18, 2016b, 93) "the discovery of the 'environment' took place in the trenches of the First World War." This discovery in the land, air, water and fire was hardly a happy one, neither for 'the environment,' nor the body.

During WWI the bodies of male soldiers and female civilians were sites of violence, including sexual violence with male perpetrators and female victims, portrayed horrifically in Louis-Ferdinand Céline's recently rediscovered and obscene *War* (2024). It is a stark reminder that all wars are fought against women and that in warfare women are often losers as well as men. Sexual violence and martial violence go hand-in-hand; sexual violence is war by other means. *War* is not about battles and armed conflict, and only briefly and at the beginning of the book about the mud of battlefield and the horrors of the war going on at the front (Céline 2024, 1, 2, 3, 5, 11). It could have been called *Sex in Time of War*. It is also about the aftermath of battle behind the lines, including in a military hospital and lazaretto with "the rotting [and disgusting] stench of the dead bodies" and the dying, gangrenous bodies that stank (9, 20, 37) that assaults the immediate sense of smell. It lays bare and brings to attention the aftermath of industrial warfare that was supposed to stay hidden, out of sensory range and out of mind. The Anthropobscene is on full view. The operating theatre was just as much a theatre of war as air, earth and water. The sexual, wounded, dying and dead body was just as much a theatre of operations as the other theatres. Modern industrial warfare is Anthropobscene.

The transformations of the bodies, spaces and places of war had a devastating effect on them, and people's perception of them. Guillaume Apollinaire (cited in Sherry 2005, 141) writing at the beginning of WWI prophesied:

> Before the war we had only the surface
> Of the earth and the seas
> After it we'll have the depths
> Subterranean and aerial space.

In fact, they had the subterranean depths and aerial heights *during* it. The spaces and places of world warfare not only had an effect on space and people's perception of it, but also on people's perception on time. Both spaces and places cut off people from their pasts by eradicating the visible signs of the persistence of the past in the present. They cast people adrift onto a sea of time disconnected from the past, uncertain of the future, and fearful in the present. Both World Wars thus not only had an effect on space, but also on time, not only then but also now for the survivors and witnesses such as Knell (2003, 2) for whom "there is a psychological need to forget and a moral obligation to remember. There is the human desire to forgive and the ethical necessity to warn of a possible repeat of the disaster."

The transformation in both wars of the natural and cultural environments in the country and the city (and in their criss-cross discussed in the final chapter) and shifts in the conduct of industrialised world warfare are indicative of a change in the conduct of war at the fundamental level of the four elements of earth, air, fire and water. It was also often remarked upon, and figured, in these terms by observers of, and commentators on, both wars.

In the case of the WWI, with its heavy artillery and trench warfare, earth and water were mixed to produce what commentators and caption writers called "swamps." The air of WWI became toxic and the landscape of trench warfare was slimy, the mixing of earth and water (in mud). War across the surface of the land in the two dimensions of length and breadth was extended into the third dimension of the heights above the land and the depths below the land.

In the case of WWII, with its aerial bombing and city warfare, fire and air were mixed to produce what commentators and historians called "firestorms." The landscape of strategic bombing in WWII was sublime, the mixing of fire and air (in exploding bombs and firestorms), and the mixing of earth and air (in destroyed cities). Fire in the sense of explosion and combustion was the agent that transformed the solid matter of cities into piles of rubble. War was fought across the surface of the land in the two dimensions of length and breadth extended further into the third dimension of the heights above the land and the depths below the land.

The sublime city ruins of WWII transcended and sublimated (in two senses) the slimy hell-holes of WWI. The horrors of the mud and slime of trench warfare were sublimated into the terrors of the fire-storms of strategic bombing. WWI in the words of who Joanna Bourke (2005, 226) calls "the great military theorist" J. F. C. Fuller was "bogged down in the slough" of trench warfare and artillery bombardment (cited by Grayling 2006, 195), whereas WWII was seared on high in the firestorms of aerial bombardment (as we will see in the following chapter). The "hallmark" of WWI was "strategic stalemate on the western front" between combatants, whereas strategic bombing of the home front of WWII meant that "military violence against non-combatants had become the hallmark of total war" (Chickering and Förster 2005, 6, 11; see also Overy 2005).

Subsequent wars, such as the Cold War of the 1950s to 1980s and the Gulf War of 1991, sublimated warfare even more from the "massive urban destruction" (Grayling 2006, 176) of WWII and the morassive underground devastation of WWI into the nightmare of 'mutually assured destruction' by nuclear weaponry and into the heights of extraterrestrial space where the satellites orbit (see Giblett 2008b, chapter 9). It is as if the horrors of the WWI could not be repeated on the surface and in the depths of the earth, but in order to avoid them new terrors were invented in the heights above the earth. In the drive to make war more clinical, distanced and remote, war was taken to new heights and greater lengths, or distances. WWI was "killing at a distance" (Carlyon 2006, 33) with artillery bombardment; WWII was killing at greater distance with aerial bombardment; and the Gulf War of 1991, at even greater distance with satellite-guided and targeted bombardment (see Giblett 2008b, chapter 9).

This drive has been going on for some time. The American Civil War was arguably the first modern industrial war as it involved the use of railways and telegraphy, two sublime communication technologies (see Giblett 2008b). For the American Civil War General William Sherman (1990, 343–344) the scene of the battle of Vicksburg was "truly sublime" and "a picture of the terrible not often seen." At the battle of Chattanooga he could not help stopping to look across "that vast field of battle to admire its sublimity" despite, or perhaps because of, the "carnage and noise" (390). Part of the sublimity of the battlefield resides in the destructive power of modern warfare. Sherman tersely stated in a letter to an editor of a newspaper that "generally war is destruction and nothing else" (301) and in another to the Mayor of Atlanta that "war is cruelty and you cannot refine it" (601). No doubt the ordinary infantryman experienced more of the destruction and cruelty of the heat of battle first hand and not much of the sublimity of the scene of battle viewed from afar by the generals. This was certainly the case in WWI: "War was return of earth to ugly earth,/War was foundering of sublimities," as Robert Graves (Walter 2006, 264) put it at the time. WWI was the foundering of sublimities in slime. It was, in a word, a portmanteau word, s(ub)lime in which, as Sofoulis (the deviser of the portmanteau) argues, slime is the secret of the sublime (as we saw in Chap. 3).

From the American Civil War, the first war of the industrial age, to WWI, "the first big war fought between industrialised societies" (Carlyon 2006, 761–762), a shift occurs in the type of warfare being conducted, not just in the scale of industrial technologies brought to bear in and on it. A mobile war fought through and over several states in the case of the American Civil War gives way to a war 'bogged down' *in* (not just on) the Western Front, including the "swampland" of Passchendaele (Fussell 2000, 84; see the map 'Passchendaele: The mud' in Gilbert 1994, 103) which depicts appropriately "waterlogged areas of the battlefield, in which the mud was deepest" as black blobs). Battles were fought in and over the same ground several times in the case of WWI. Artillery bombardment created craters and troops dug trenches to escape this and other firepower. Rain filled both craters and trenches producing mud, a persistent feature of the landscape of trench warfare.

7.2 Mud

Those who were there and wrote of it, such as Edmund Blunden, Frederic Manning, Frank Hurley, Edward Lynch, Edward Thomas and Louis-Ferdinand Céline, refer to mud repeatedly and use the word persistently.

A later commentator in Ellis (1976, 43–47) devotes several pages to the discussion of mud in WWI, and even about drowning in it. Fussell (2000, 16) states that at the battle of Passchendaele in 1917 "thousands literally drowned in the mud." For Helphand (2006, 27–28) too "the land of Flanders became a foul, seething swamp" in which "thousands drowned in the mud." More precisely, the land was turned into artificial wetland. In the aftermath of the war for Siegfried Sassoon (cited by Lloyd 1998, 129) "the Dead [...] struggled in the slime." In death as in life there was no escaping from slime and mud. The Dead lived on in slime. In Lynch's *Somme mud* published for the first time in 2006 what he calls "our world of mud" (31) is a 'manmade' environment that has replaced the earth:

> We live in a world of Somme mud. We sleep in it, work in it, wade in it and many of us die in it. We see it, feel it, eat it and curse it, but we can't escape it, not even by dying. (Lynch 2006, 108)

Mud commands the processes of living and dying, the men who are living and dying, and times of the present and future (dead or alive) invoking the apocalyptic: "the reeling, rocking earth seems to be toppling to its very doom" (58). Besides addressing the senses of seeing, feeling and tasting, it addresses the senses of hearing ("a splashy oozy sound" (112)) and smell ("foul-smelling" (108)). It attacks and attaches to the surface of the body ("muddied skin" (85) and "mud-stained [...] as the place itself" (108)). It crosses the border between outside and inside.

Robert Graves' *Goodbye to all that* mentions mud occasionally, though generally he prefers the more colourful, onomatopoeic and tactile, "morass." Outdoing the Australians Hurley and Lynch and the Englishmen Blunden and Graves for colour and drama, though, is the Frenchman Henri Barbusse. The very first chapter of his classic *Under fire* sets the scene for "this war of mud":

> In the sinister light of the storm beneath black dishevelled clouds, dragged and spread across the earth like wicked angels, they seem to see a great livid white plain extend before them. In their vision, figures rise up out of the plain, which is composed of mud and water, and clutch at the surface of the ground, blinked and crushed with mire, like survivors from some monstrous shipwreck. (Barbusse 2004, 6)

In this apocalyptic vision of the war of the worlds "under attack from the elements" (153) earth and water are mixed up chaotically in mud and mire.

From this conclusion to the first chapter entitled 'The Vision,' *Under fire* segues into the second chapter entitled "*in* the ground" (my emphasis). This chapter begins by taking the bird's eye aerial view of:

> The earth! The desert has started to appear, vast and full of water, beneath the long and desolate light of dawn. Ponds and craters, their waters pinched and shivering under the sharp blast of early morning; tracks left by last night's troops and convoys in these fields of sterility, streaked with ruts which shine like steel rails in the weak light; and piles of mud with here and there broken stakes emerging from them, broken trestles in the shape of a cross, packages of rusted wire, twisted, like bushes. With its puddles and its banks of slime it looks like an oversized grey canvas sheet floating on the sea, submerged in places. It is not raining, but everything is wet, oozing, sodden, drenched. The pale light seems to be pouring across it. (7)

By mixing earth and water, slime is produced, an unsettling substance that is neither solid nor liquid, but the solid becoming liquid (as we saw in Chap. 3). In the lexicon of trench warfare "slimy" means "abandoned by water and by men" (138). The slimy also means abandoned by God. This is what he calls in Biblical terms "the diluvian scene" (8), the world of the flood punished by God.

Modern industrial war turns the earth into a wet desert in the sense that this is largely lifeless land—it is deserted. A deserted trench is "a frightful, horizontal desert" (185). It is not the horizontal desert of a living ecosystem home to native species but "the waste lands [...] of the sodden filth that is pressed/Into ooze, and the sombre dirt that spouts up" as D. H. Lawrence put it (Walter 2006, 28). It is a manmade unhome of "this dead place," to return to Barbusse (2004, 189), of the human species "buried in molehills" (17) in "this tomb" (233). It is a "wasteland, dirty and sick, [...] mired with black mud" on which "the dead are lined up" (136). "Space has shrunk and the field of sadness of the sky has clamped down on the earth, that field of death" (25). The earth has been transformed from a cradle of life into a field of death, from womb to tomb. The landscape of trench warfare is the dead black water and earth of an artificial, 'man-made' wetland. It is the quintessential landscape of the Anthropobscene in several senses of being manmade and laying bare the havoc of industrial warfare, its waste of energy and its devastation of the earth. It is "swampy waste ground" (139), not the living black waters of a wetland (see Giblett 1996). It is an artificial quaking zone, not the "trembling earth" of a swamp, such as (literally) Okefenokee in Georgia, USA (Nelson 2005). The wet nether land of trench warfare is an artificial grotesque lower earthly stratum (see Giblett 1996). To these artificial wetlands the same horror is attached that is often associated with the wet nether lands of swamps, marshes and bogs.

Modern industrial war also transforms the earth into the moon in the sense that the land has become a cratered landscape, or more precisely "marshy craters" (Barbusse 2004, 298) for it has been flooded. The landscape of trench warfare is "like a lunar landscape" for Erich Maria Remarque (1996, 40) in his WWI classic *All Quiet on the Western Front*. For Frederic Manning (1999, 160) too it is "some sterile landscape in the moon." For Wilfred Owen (cited by Helphand 2006, 31) "No Man's Land under snow is like the face of the moon chaotic, crater-ridden, uninhabitable." Similarly for Winston Groom (cited by Helphand 2006, 34) no-man's land is "the barren and ruined moonscape." The earth has become the moon because it has been transformed into a barren place that is unable to give birth to new life, unlike fecund wetlands, one of the most productive ecosystems and habitats on earth (see Giblett 1996).

The landscape of aerial bombardment is also a lunar landscape of craters. This was the case with WWII, though the craters were dry unlike the wet craters of WWI, and of the Vietnam War. This was a wetland war fought partly in and over the wetlands of the Plain of Reeds and the Mekong Delta, the most fertile zone of Vietnam (see Giblett 1996, chapter 9). After seven years of bombardment, over 20 million craters spread over 300,000 acres in Vietnam made this war, as Helphand (2006, 17) puts it, "as much against the land as against armies." The spraying of Agent Orange to reduce cover provided by foliage also demonstrates that this was a war fought against the land and its inhabitants. Yet not only the Vietnam War but all modern industrial wars were wars fought against the land as WWI and II attest. The Cold War was also fought against the land with its nuclear bomb-testing in the Australian and Nevada deserts and elsewhere on and under the earth and sea. Yet not only modern industrial wars are fought against the land; modern industrial warfare is unofficially declared against the land. Other undeclared wars are fought against it too. Modern industrial chemistry for Rachel Carson (1962, 7) in her classic

Silent spring is a war against nature (see Giblett 2008a, chapter 6). The same could be said for modern military physics.

The moonscape is a dead landscape in which, rather than choirs of angels singing men to their deaths, "diabolical noise surrounds us" (Barbusse 2004, 193) in "this landscape for witches' sabbaths" (206). In this world of fallen angels a black mass transubstantiates in an anti-communion the sacred elements of earth and water into the manmade mud and slime of the landscape of trench warfare. This is not the water-land of a fen, marsh or swamp, nor the pre-diluvian creation scene when God created the heavens and the earth, and had not yet divided the water and the land. Rather, it is the world of the flood in which God punishes sinners (but who is Noah et al saved in his Ark?). All that is solid melts into liquid. The hard becomes soft; the solid object becomes the slimy abject. Water brings about "the softening of the earth" (153). What was hard becomes soft.

Mud not only attacks, and attaches to, the surfaces of the body and clothing but also becomes an all-encompassing, all-enveloping medium into which the body enters. At one point the first-person narrator sees his comrades "slowly sliding, rolling in the mud in the depth of the night." They "sank as far as we could into the mud" and "pressed themselves further and further into the ground" (208). They end up "buried by night." The first-person narrator is able later to "emerge from the earth in which we have been encased" (211). They have been "dirtied by their contact with the earth, dirtied by the grey smoke of space" (212). As dirt for Mary Douglas (1966) is "matter out of place," body and earth have ceased to be separate but have been mixed together. Earth outside has come inside. Indeed, they are not only consumed by the mud, but "almost digested by the earth" (Barbusse 2004, 228). The earth becomes an orally sadistic monster:

> As you sink into the depths a pestilential atmosphere, heavy as the earth itself surrounds you. Your hand feels the clay wall; cold, sticky, sepulchral. This earth weighs on you from all sides, wrapping you in dismal solitude, touching your face with its blind, mouldy breath. (256)

The landscape of trench warfare is not the great dismal swamp of a living wetland but the grimy damp 'swamp' of a dying wasteland.

In contrast to Barbusse, the earth for Erich Maria Remarque (1996, 39) was both a nurturing mother and an object of oral sadism:

> The power to defend ourselves flows back into us out of the earth and out of the air – and most of all it flows out of the earth. The earth is more important to the soldier than to anybody else. When he presses himself to the earth, long and violently, when he urges himself deep into it with his face and with his limbs, under fire and with the fear of death upon him, then the earth is his only friend, his brother, his mother, he groans out his terror and screams into its silence and safety, the earth absorbs it all and gives him another ten seconds of life, ten seconds to run, then takes hold of him again – sometimes for ever. Earth with your ridges and holes and hollows into which a man can throw himself, where a man can hide! Earth- in the agony of terror, the explosion of annihilation, in the death-roar of the shell-bursts you gave us that massive resurgence of reconquered life. The madness, the tempest of an existence that had practically been torn to shreds flowed back from you into our hands, and so we burrowed deep into you for safety, and in the speechless fear and relief of having survived the moment, our mouths bit deeply into you!

Remarque's soldier is the subject of oral sadism who bites the earth before the earth bites him back as it does for Barbusse's soldier. Remarque phantasizes a benevolent mother earth and hardly mentions "the black mud" (47) that blots the landscape and transforms the earth from something dry and solid into something wet and liquid.

The horror of the slimy for Barbusse (2004, 7; see also 9) has a particular bodily inflection (and infection) and a specific sensory address to the sense of smell:

> You can see a maze of long ditches in which the last remnants of night linger. This is the trench. The bottom of it is carpeted with a viscous layer that clings noisily to the foot at every step and smells foul around each dugout because of the night's urine. The holes themselves, if you lean over them as you pass, smell like a whiff of bad breath.

Like the natural wetland, "the soft, sticky mud" (153) of the artificial wetland attacks (and attaches to) the surfaces of the body and clothing. Mud has agency as "we waded through marshy field, we got lost in silent regions where the mud grabbed us by the feet" (189). Slime even penetrates inside the body via the sense of smell when "a smell of dank cellar and marshland penetrates us" (283).

However, unlike its natural counterpart in which the smells are produced by the non-human and by the processes of non-human decomposition and recomposition, the smells of the artificial wetland of the trench are produced by human waste out of which new life springs (eventually). One soldier says "the place smelt of shit" (49). This is hardly surprising when they were in "the bowels of the earth" (122, 193, 233; see also Isaac Rosenberg's poem 'Break of Day in the Trenches' (cited by Fussell 2000, 250–251—for him "the greatest poem of the war"—and in Walter 2006, 48). Yet this is not the cloaca of a wetland where defecation and procreation are mixed in the grotesque lower earthly stratum and from which new life springs. Rather it is the manmade sewer of the nether lands of the grotesque artificial lower earthly strata.

The manmade atmosphere of trench warfare was not only smells emanating from the ground below, but also gas wafting in the air above ground. It could spell death but also create a psychopathology of smell. "The fear of gas obsessed" Graves (1929, 278) to such an extent that "any unusual smell, even a sudden strong scent of flowers in a garden, was enough to send me trembling." The sense of smell becomes heightened in the feral quaking zone of trench warfare, as does the sense of touch. Through the soles of his feet and boots it was possible for Barbusse (2004, 286) to distinguish between mud and shit when "we are walking though filth and can feel its slithering softness against the earthy mud." "The mire of excrement" and "the foul and fetid muck" (287) is different from the rich and fertile mud of a living wetland. They are often confused, though, partly due to the indiscriminate sense of smell which tends to associate all smells of decomposition with shit as death-dealing waste and not as life-giving manure.

Trench warfare for Barbusse is a human zoo in which "we are drenched in the odour of a lion's den" (56). Despite the fecund smells, this is a sterile place, "the grey, muddy, downtrodden landscape in front of us [is] sterilized by the rain" (109). The water of the battlefield is not life-giving, nor even death-dealing, just sterile with no possibility of new life springing from it. The heat of thermodynamic industrial warfare has scarred the living body of the earth producing "the huge sterilized, cauterized plain" (139). A scorched earth policy of burning has been enacted on the battlefield itself, not merely on the pasture and crop fields of retreating armies for invading ones. This wounding has also penetrated beneath the surface. Not only has the surface of the body of the earth been cauterised and scarred but also the depths of the body of the earth have been exposed with "giant viscera and entrails opened upon the earth" (207) leaving "the plain with its entrails bleeding and rotting" (270). The murky depths have become the open surface, the body of the earth opened up.

WWI not only had its physical effects on men and landscapes; it also produced psychological affects such as the "black depression" of what the French called "*le cafard*, the cockroach, crawling up their backs" (Winter 2004, ix), or more simply in African-American terms, "the blues" (Helphand 2006, 45; see also 116). This black depression under the black sun of melancholy affects both the soldier and the earth as what Barbusse (2004, 205) calls "a dreadful melancholy crushes the earth." The earth is burdened not only with destruction, but also with depression, just like its inhabitants. There is a parasympathetic response in humans and the earth. This is an unhealthy place for mental health. These are not the old fears of "the miasma of the slums" (17) (as we will see in Chap. 9) and the swamps of sickness in which malaria arises (as we saw in Chap. 2). Rather it is the melancholic marshes of "a lake of stagnant darkness" (207) on whose "pestilential shore" (248) horror and terror, the two ruling passions of modernity, arise. This is "a zone of terror" (298) of "soft ruins" and "dreadful space." Terror of the world above the ground, horror of the world below; terror of guns, mortar, shells, planes and zeppelins; horror of slime, mud, drowning, death, decomposition. More precisely, the world below is where "the bodies [are] buried under the mud of the earthly lower depths" (314).

Otto Dix graphically illustrated this obscenery in which body and land merge together in destructive association. Skeletons and skulls stuck in the mud are revealed when trenches collapse (see Henshaw and Horsfield 2007, plate 9 'Collapsed Trenches,' 3). Body and land are no longer the benign figure for each other (see Giblett 2008a). Nor is landscape simply the backdrop for the scenery of war. The landscape for Henshaw and Horsfield (2007, 2) in Dix's prints is "torn apart, desecrated and ravaged. Often the landscape appears alien, other-worldly, nightmarish. It appears sometimes as a simple backdrop to human tragedy, but often as a more integral part of the destruction" as in 'Collapsed Trenches' that depicts what they call "disarticulated limbs." Land and body merge together in obscene and destructive embrace. Bits of bodies are parts of the landscape. Paul Nash also illustrated these 'trench-scapes,' or "antilandscapes" as Samuel Hynes (cited by Helphand 2006, 36) calls them without the graphic detail of Dix's drawings, yet depicting the transformation of the land from beneficent landscape to malign anti-landscape.

The land for Wilfred Owen (cited by Helphand 2006, 32) is not only "cratered like the moon" but also "pitted with great pocks and scabs of plagues." Commenting on this figuration of the land as body, Helphand (2006, 32) goes on to argue that

> in anthropomorphizing the landscape as a body, writers described the face of the land as deformed and disfigured and the body of the land as racked by disease and plague, ridden by the pestilence of war. Unlike the soldiers who died, the landscape survived, but it too was a casualty of war.

Body and land have become the malignant figure for each other; both are diseased not only physically but also psychologically, both are obscene.

The world of men above is mirrored for Barbusse (2004, 318) in the world below:

> You can see their shadows moving across the great melancholy mirror of the ground and reflected in the pale stagnant surface of the old trenches, which alone whitens and inhabits the infinite void of space in the midst of this polar desert with its misty horizons.

Rather than the reflective surface of water being a mirror in which soldier could reflect upon his place in the scheme of things, the surface of water reflects the sterile desert of the world above. The solid became liquid, in a word, slime, but the solid could also become gaseous, ethereal, in a word, sublime. For Blunden (2000, 108) "the solid ground was ethereal" when they felt that they were able to rise above it, exultant, sublime. Whilst the world above ground, such as the sky, could evoke the sublime, even in war (see Fussell 2000, 52–53), the world below was the impure realm of the uncanny—in war, or not. The landscape of battle in WWI for Winter (2004, xv) was "full of the uncanny and the bizarre." Charles Carrington (cited by Fussell 2000, 79) was obsessed for 50 years with "No Man's Land" and "the unknown world beyond it [...O]ver there [...] is the uncanny." The uncanny is elsewhere, not here, not now, but always somewhere else, lurking and malevolent.

The landscape of trench warfare is not only a landscape of space (and place) but also of time in which the future is colonised. The soldiers of trench warfare are not only buried in mud but also, as Barbusse (2004, 8) put it, "we are buried in the depths of an eternal battlefield" in which the past stretches out behind them and the future rises before them. Time has become space; time has been spatialised. They have looked back into the past only to see the future with horror for "he has seen too deep into the past not to see the future with horrifying precision" (127). They are avatars of Walter Benjamin's and Paul Klee's *Angelus Novus* who faces, and looks back into, the past whilst moving backward into the future, the future of WWII (as we will see on several counts in the following chapter).

7.3 Hell

Understandably in the struggle to explain the inexplicable Barbusse invokes hell. William Sherman had already stated in the mid 1860s that "war is hell." The Prussian Chancellor, Otto von Bismarck (cited by Sloterdijk 1987, 219), said the same about Franco-Prussian War later in the early 1870s. The poet-painter, William Blake, in the eighteenth century had already critiqued that "war is energy enslaved." As Sloterdijk (2018, 10) says, "wars represent the worst form of the waste of energy." The fossil fuels of solid coal and liquid oil are sublimed into steam and gas to fire-up and power the machines of modern industrial warfare that turn the earth into watery hell, into the Anthropobscene. For Barbusse (2004), the "hell" (47), "this hell on earth" (209), and "this hell" (257) of trench warfare is not so much "a blazing inferno with huge and furious masses of red and black" (226) as ultimately, and in a word, water:

> At one time I thought that worst hell of war was the flaming of shells; then, for a long time, I thought it was the suffocation of underground passages that constantly close in on you. But no: hell is water. (299)

Barbusse's narrator does not go on to say why, and presumably there is no need to do so given all the reasons that have already been enunciated.

The watery wasteland of WWI trench warfare for Barbusse (2004, 306) mixes the elements of earth and water:

> I shall never forget the sight of this endless landscape across the face of which the filthy water had corroded colours, features, relief, its forms attacked by this liquid decay crumbling and draining off on all sides through the broken skeleton of posts, barbed wire and shoring – and above it all, amid the dark Stygian immensities, the vision of this thrill of reason and simple logic which suddenly began to shake these men like a fever of madness.

The ancient river Styx of the underworld is no longer an infernal, imaginary marshland as it was for Dante (*Inferno*, VII, 103–130) in which the souls of the sullen are stuck (as we saw in Chap. 3), but has become an artificial, all-too-real marshland, "this sullen mile of mud," as A. P. Herbert (cited by Lloyd 1998, 25) put it, in which sullen soldiers are stuck. The underworld has also ceased to be down below and come to the surface to be revealed in all its horror, truly hell *on* earth in a revelation of the Anthropobscene. This is a particularly and peculiarly modern hell in which the land has been transformed by industrial technology. Ypres for one soldier was "a proper Dante's inferno. Ground churned to pulp and mire" (cited by Helphand 2006, 32).

The hell of the battlefield becomes the modern landscape of hell par excellence. Indeed, for Ernst Bloch (1998, 103) writing in 1931 "in modernity's hell there exists an emptiness that is equivalent to the emptiness of the modern battlefield." The hell of the modern battlefield is not merely a 'symbol' for the hell of modernity, but is contiguous with it. Writing two years earlier, Bloch (1998, 308) was unfortunately prescient about this contiguity as he suggested that "the nothingness that stands behind the mechanized world, a world unmediated by humanity, is a mortuary in which people have been buried alive." In fact, in WWII the bombed cites of Europe were mortuaries in which people were buried alive in baleful fulfilment of Bloch's prophecy.

Views of the modern battlefield as a muddy hell are not the monopoly of eyewitness accounts but persist in a recent historical novel of WWI by Jeff Shaara (2004) in which "the ever-present mud, the soggy lowlands of Flanders" is "a black, wet hell" (3) with "black water" (4, 6, 8 and 9). This is a hell of smell emitted by "unseen decay" from "what lay rotting in the deep mud" (5). As for Lynch and other writers, mud is the dominant substance and word for Shaara (6). This hell *on* and *in* earth with the mixing of earth and water in WWI becomes the hell *above* earth in WWII in which fire and air are mixed in the firebombing of aerial warfare, a hell of flames *on* earth with the destruction of cities (as we will see in the following chapter).

The muddy, slimy landscape of WWI was also well-documented and dramatized by war photographers. The official Australian war photographer Frank Hurley did so on many occasions. He is most famous for his composite photograph comprising two of his own photographs, one of a sunburst through clouds and the other of a battlefield. The composite is called 'Morning after the Battle of Passchendaele or An Advanced Aid

Post' (Willis 1988, 185, Figure 114; Lakin 2008, 60). It is arguably the most famous and most controversial Australian war photograph ever produced (for the two images that make up this composite, see Jolly 1999, figs 3 and 4). The composite, and not the battlefield scene alone in the bottom half of the photograph, made it into the official history of Australian war photography (Lakin 2008, 60), whereas the battlefield scene, and not the composite with its sunburst, made it into the official history of the war as Charles Bean (1923, vii and Figure 403), the official Australian war historian, distinguished between "photographs for the historical record, and the taking of them for propaganda or for the press, [which] were to some extent conflicting activities. Captain Hurley devoted himself to the latter work."

Hurley cast himself in the grander role of the purveyor of nationalistic sentiment. His devotion to this work was not without conflict with Bean and "Authorities" as Hurley (1917, 52 and 58–59; see also Jolly 1999) records in his diary that he:

> had a great argument with Bean about combination pictures. [I] am thoroughly convinced that it is impossible to secure effects, without resorting to composite pictures [...] Our Authorities will not permit me to pose any pictures or indulge in any original means to secure them. They will not allow composite printing of any description, even though such be accurately titled, nor will they permit clouds to be inserted in a picture.

Perhaps with good reason as Hurley's composite, with what Willis (1988, 184) calls "its sublime backlit clouds rising over a scene of death and devastation," aestheticises the landscape of trench warfare, or more precisely, the artificial wetland and feral quaking zone of the Western Front. For Hurley, as Helen Ennis (2004, 143) argues, "the battlefield was sublime," perhaps the last thing "Authorities" wanted. The battlefield was indeed included in landscape and recruited for the aesthetic under the sign of the sublime.

Yet rather than a field of battle, this scene is a depiction after the event of what Willis (1988, 187) calls "fields of mud," "a flattened, rubble-strewn, burnt-out, and ravaged landscape." This landscape of destruction is the mud and slime of the feral quaking zone of the wetwasteland of trench warfare commented upon so often by WWI writers as we have seen, and photographed by Australian photographers too (and described by Hurley in his diary). See, for example, 'The swamps of Zonnebeke,' by an unknown photographer (Willis 1988, Figure 9 and Bean 1923, Figure 400), plus other 'swamps' (Bean 1923, Figures 372, 394, 396), including Chateau Wood also in Lakin (2008, 78–79) and now ascribed to Wilkins (Maddern 2021, 21, 27). By contrast, in Hurley's composite photograph of Passchendaele the sublime and the slime counterpose each other in one composition and show their complementarity encapsulated in the portmanteau word s(ub)lime in which, as Sofoulis (as the deviser of the portmanteau) argues, slime is the secret of the sublime (as we saw in Chap. 3). In Hurley's composite the slimy land dominates the photo as it takes up two thirds of its area (and so conforms to the pictorialist rule of thirds; see Willis 1988, 142), but the slimy land is surmounted and transcended by the sublime sky in the top third. The overall effect for Willis (1988, 184) of this and other photos by Hurley is "to glorify war." The slime and mud of the depths of the earth and the wastewetland of the hell and gore of trench warfare below are sublimated into the heavens and glory of heroism above. Slime is not just the secret of the sublime, but its 'dirty secret', the uncanny counter to the landscapes of the sublime.

These 'swamps' were more precisely, artificial wetlands made up of mud, slime and skeleton trees in the feral quaking zone of the battlefields of the Western Front. 'Artificial wetlands' are usually thought of as newly created arrangements of land, water and plants designed and constructed to rehabilitate a degraded wetland or to simulate elsewhere the form and functions of a native wetland. The 'swamps' of the Western front were artificial wetlands in the sense of being newly created arrangements of land, water and plants made by destroying wetlands or by creating them where they had not existed previously with the assistance of high rainfall. The former emphasises the creative, the latter, the destructive, as did the landscape of WWII. The reshaping of the arrangements between the four elements of earth, air, fire and water took place too in the case of WWII with its long-range bombers. Air and fire were mixed to produce firestorms destroying cities and turning them into artificial ruins and many of their residents into charred remains in the feral quaking zone of the bombed cities of the home front.

7.4 Technology, the Body, and Benjamin

This reshaping of the land, in turn, had its effect on those who fought these wars and were affected by them. The effects of this reshaping of the land on people has not often been commented upon, except with the notable and remarkable exception of Walter Benjamin (1999, 732) who argued that in what he called "positional warfare" of WWI:

> a generation that had gone to school in horse-drawn streetcars now stood in the open air, amid a landscape in which nothing was the same except the clouds and, at its centre, in a force field of destructive torrents and explosions, the tiny, fragile human body.

Benjamin brings together philosophy, ecoculture and psychopolitical ecology in an exemplary manner (as we saw in Chap. 3). Sadly, tragically, unfortunately (it is impossible to avoid clichés and to find adequate words to describe this loss, his loss) Benjamin took his own life in 1940 before he could remark about the next generation that had gone to school in electric streetcars (trams) and steam railways who now in the strategic warfare of WWII sat huddled in the closed air of the bomb shelters amid a landscape of ruins in which they could not even see the clouds, but still at its centre in a force field of greater destructive torrents of fire and exploding bombs, was the still tiny, fragile human body. Yet whereas the tiny, fragile human body that Benjamin remarked upon in WWI trench warfare of the country was the soldier body, the tiny, fragile human body in WWII strategic bombing of the city was the civilian body. Whereas the landscape of trench warfare in WWI was the landscape of German idealism, the landscape of bombing warfare in WWII was the landscape of German fascism and its allied counter-force producing the conflagration of cities and citi-zens. Both were anthropobscene.

Benjamin (1979, 103–104 and 1996, 486) noted in *One-Way Street* the beginning around the time of WWI of the conditions of technologically advanced modernity under which:

> human multitudes, gases, electrical forces were hurled into the open country, high frequency currents coursed through the landscape, new constellations rose in the sky, aerial space and ocean depths thundered with propellers, and everywhere sacrificial shafts were dug in Mother Earth. This immense wooing of the cosmos was enacted for the first time on a planetary scale, that is, in the spirit of technology.

Benjamin (1979, 104, 1996, 487) continues by bemoaning the fact that

> the mastery of nature, so the imperialists teach us, is the purpose of all technology. But who would trust a cane wielder who proclaimed the mastery of

> children by adults to be the purpose of education? Is not education above all the indispensable ordering of the relationship between generations and therefore mastery, if we are to use this term, of that relationship and not of children? And likewise, technology is not the mastery of nature but of the relation between nature and man [*sic*].

The ultimate technological mastering of nature (not of the relationship with it) for Benjamin (2003, 393) is technocracy (the rule of technology, the rule of technocrats) that emerged in fascism. Benjamin found the roots of the technological monstrous in what he called "machine warfare" of WWI in which "everything sober, unblemished, and naïve has been considered regarding the improvement of human society ends up between the worn teeth of these Molochs, who react with the belches of 42-cm. mortars" (Benjamin 1999, 319). Benjamin (1999, 319) concludes that the "old mystery of nature, far from revealing itself to their technology, is exposing its most threatening feature. In the parallelogram of forces formed by these two – nature and nation – war is the diagonal." All wars are wars against nature, or living beings and matter.

For Benjamin (1979, 99, 1996, 483) in order "to turn the threatening future into a fulfilled "now," the only desirable telepathic miracle, is a work of bodily presence of mind." For Benjamin the body is, or can be, in the fulfilled present in the mind and in the now going into the future. This was certainly the case for Benjamin (1979, 99, 1996, 483) as he goes on to suggest that "primitive epochs, when such demeanor was part of man's daily husbandry, provided him, in the naked body, with the most reliable instrument of divination [...] What would have become a portent of disaster he binds bodily to the moment, making himself the factotum of his body." "Primitive man" does all kind of work *for* his body (dance, ritual, tai chi, yoga, etc.), whereas "modern man" does all kinds of work *with*, or *on*, his or her body (body-building, diet, exercise, etc.). The traditional human being works his or her whole body and not just parts of it. Rather than the body being a portent of disaster in the future (of disease, decay and death), the body sutures him/her to the present and fulfills the now and the here.

Benjamin later commented on "modern man" and his relationship to his own body. Writing on the tenth anniversary of Franz Kafka's death, Benjamin (1973, 126, 1999, 806) argues that in Kafka's *The Castle*, "just as K lives in the village on Castle Hill, modern man lives in his body; the body slips away from him, is hostile towards him [...] Exile – his exile – has gained control over him[...] the most forgotten alien land is one's own body." Benjamin lived this exile from his own body, and a double exile from his own homeland. By contrast, for traditional human beings the most fondly remembered homeland is one's own body. Exile is not just a matter of leaving one's home and home-land, but is also experienced in relation to one's own body, something that Benjamin wrestled with his whole life. K is not at home in his own body; the castle is not a home. Being at home includes being at home in one's own body and place. For traditional humans, one's land is one's own body, and vice versa

7.5 Melancholy in the Aftermath

The muddy landscape of WWI had a devastating psychological effect afterwards on those who were not there. Vera Brittain's account of her visit to Venice in the aftermath of WWI in her autobiographical study, *Testament of Youth*, plays out the association of melancholy and mourning with the city of Venice. As she "glides smoothly" in a gondola "over the rippling grey silk of the Grand Canal" on her post-WWI tour of Europe, the view of Venice for Brittain (1978, 479) is tinged with melancholy and mourning as she associates the Venetian waters with the death of her fiancé in France and brother in Italy during the War. The Grand Canal is associated not only with melancholy and mourning, but also with magic as Brittain "with melancholy possessiveness [...] looked upon those enchanted waters... those fairy lagoons, incredible as a gorgeous mirage in the muffled silence." These are the benign waters of the patriarchal mother earth. She imagines that her brother "had died saving this beauty from the fate of Ypres," in other words, saving the beauty of the city of Venice from the horror of trench warfare, especially its mud.

Brittain read about the mud of trench warfare in the letters her brother and fiancé wrote to her and she had immediate experience of it as a nurse in France and in the aftermath of the death of her fiancé. In one of his letters from the trenches Brittain's fiancé related that they were "very wet and muddy [...] The whole of one's world, at least one's visible and palpable world, is mud in various stages of solidity or stickiness" (Brittain 1978, 206). Mud mixes the elements earth and water, whereas cities and roads separate them as a rule, or should do. Brittain (1978, 337) experienced the mud firsthand when she arrives on the western front and found that "the roads were liquid with such mud as only wartime France could produce after a few days of rain." The solidity of earth was liquefied into mud.

The mud of trench warfare, especially its smell, in associated for Brittain with death when she goes through the "kit" of her fiancé following his death. She wrote to her brother that everything was "simply caked with mud" and she felt "overwhelmed by the horror of war" as "the smell of those clothes was the smell of graveyards and the Dead. The mud of France which covered them was not ordinary mud; it had not the usual clean pure smell of earth, but it was as though it were saturated with dead bodies – dead that had been dead a long, long time" (Brittain 1978, 225). The mud of trench warfare had the smell of death in which earth, water and dead bodies (and shit (though she does not mention this)) were mixed. It did have not the usual smell of mud in which earth and water were mixed. Brittain distinguishes the clean, pure smell of earth from the horrific, dirty smell of the mud of trench warfare.

Like most writers about WWI (as we have seen in this chapter), Brittain (1978, 355) associates the mud of trench warfare with swamps and marshes, such as when she describes how "the terrific gales and whipping rains of late autumn [...] turned the shell-gashed flats of Flanders into an ocean of marshy mud that made death by drowning almost as difficult to avoid as death from gun-fire." Yet the wetlandscape of trench warfare was an artificial marsh made by modern industrial warfare, not a native marsh made by hydrogeological processes and ancestral hands. It was a feral quaking zone, not a native quaking zone.

The wetlandscape of trench warfare is also melancholic for Brittain (1978, 356), such as when she describes how "the Flanders offensive was subsiding dismally into the mud" and refers to "melancholy Flanders" on the next page. Brittain experiences mourning and melancholia during and after WWI, yet she also experiences these affects in Venice after the war in combination with the magic of the maternal waters of the wetlands—perhaps hardly surprising given her strong feminist beliefs, though she is unable to distinguish the canalized waters of the patriarchal city from the matrifocal waters of the wetlands that preceded the city and which the city destroyed to be built. Although she distinguishes the dirty impure smell of trench warfare from the clean pure smell of earth, she does not regard the former as the product of patriarchal and industrial trench warfare and the latter as the progeny of maternal marshes, nor does she distinguish the artificial 'manmade' marshes of modern industrial warfare from the native marshes made by hydrogeological processes and ancestral hands.

References

Barbusse, Henri. 2004. *Under Fire*. Translated by Robin Buss. London: Penguin. First published in French in 1916.

Bean, Charles. 1923. *The Official History of Australia in the War of 1914–1918: Volume XII, Photographic Record of the War*. Sydney: Angus & Robertson.

Benjamin, Walter. 1973. *Illuminations*. Translated by Harry Zohn. London: Fontana.

Benjamin, Walter. 1979. *One-Way Street and Other Writings*. Translated by Edmund Jephcott and Kingsley Shorter. London: NLB.

Benjamin, Walter. 1996. In *Selected Writings: Volume 1, 1913–1926*, ed. Marcus Bullock and Michael Jennings. Cambridge, MA: Belknap Press of Harvard University Press.

Benjamin, Walter. 1999. *Selected Writings: Volume 2, 1927–1934*, ed. Michael Jennings, Howard Eiland, and Gary Smith. Translated by Rodney Livingston and others. Cambridge, MA: The Belknap Press of Harvard University Press.

Benjamin, Walter. 2003. *Selected Writings: Volume 4, 1938–1940*, ed. Howard Eiland and Michael Jennings. Translated by Edmund Jephcott and Others. Cambridge, MA: The Belknap Press of Harvard University Press.

Bloch, Ernst. 1998. *Literary Essays*. Translated by Andrew Joron and Others. Stanford, CA: Stanford University Press.

Blunden, Edmund. 2000. *Undertones of War*. London: Penguin. First published 1928.

Bourke, Joanna. 2005. *Fear: A Cultural History*. London: Virago.

Brittain, Vera. 1978. *Testament of Youth: An Autobiographical Study of the Years 1900–1925*. London: Virago.

Carlyon, Les. 2006. *The Great War*. Sydney: Pan Macmillan.

Carson, Rachel. 1962. *Silent Spring*. Boston: Houghton Mifflin.

Céline, Louis-Ferdinand. 2024. *War*. Translated by Charlotte Mandell. New York: New Directions. First published in French in 2022.

Chickering, Roger, and Stig Förster. 2005. Are We There Yet? World War II and the Theory of Total War. In *A World at Total War: Global Conflicts and the Politics of Destruction, 1937–1945*, ed. Roger Chickering, Stig Förster, and Bernd Greiner, 1–16. Washington, DC: German Historical Institute/ Cambridge: Cambridge University Press.

Douglas, Mary. 1966. *Purity and Danger: An Analysis of the Concepts of Pollution and Taboo*. London: Routledge.

Ellis, John. 1976. *Eye-Deep in Hell*. London: Croom Helm.

Ennis, Helen. 2004. *Intersections: Photography, History and the National Library of Australia*. Canberra: National Library of Australia.

Friedrich, Jorg. 2006. *The Fire: The Bombing of Germany 1940–1945*. Translated by Alison Brown. New York: Columbia University Press.

Fussell, Paul. 2000. *The Great War and Modern Memory*. New York: Oxford University Press. First published 1975.

Giblett, Rod. 1996. *Postmodern Wetlands: Culture, History, Ecology*. Edinburgh: Edinburgh University Press.

Giblett, Rod. 2008a. *The Body of Nature and Culture*. Basingstoke: Palgrave Macmillan.

Giblett, Rod. 2008b. *Sublime Communication Technologies*. Basingstoke: Palgrave Macmillan.

Gilbert, Martin. 1994. *The Routledge Atlas of the First World War*. 2nd ed. London: Routledge.

Graves, Robert. 1929. *Goodbye to All That*. London: Penguin.

Grayling, A. C. 2006. *Among the Dead Cities: The History and Moral Legacy of the WWII Bombing of Civilians in Germany and Japan*. New York: Walker.

Helphand, Kenneth. 2006. *Defiant Gardens: Making Gardens in Wartime*. San Antonio, TX: Trinity University Press.

Henshaw, Mark, and G. Horsfield. 2007. *War: The Prints of Otto Dix*. Canberra: National Gallery of Australia.

Hurley, Frank. 1917. *My Diary, Official War Photographer Commonwealth Military Forces. […]*. Canberra: National Library of Australia, Ms 883.

Jolly, Martyn. 1999. Australian First-World-War Photography: Frank Hurley and Charles Bean. *History of Photography 23* (2): 141–148.

Knell, Hermann. 2003. *To Destroy a City: Strategic Bombing and its Human Consequences in World War II*. Cambridge, MA: Da Capo.

Lakin, Shaune. 2008. *Contact: Photographs from the Australian War Memorial Collection*. 2nd ed. Canberra: Australian War Memorial.

Lloyd, David. 1998. *Battlefield Tourism: Pilgrimage and the Commemoration of the Great War in Britain, Australia and Canada, 1919–1939*. Oxford: Berg.

Lowe. Keith. 2007. *Inferno: The Fiery Destruction of Hamburg, 1943*. New York: Scribner.

Lynch, E. P. F. 2006. *Somme Mud: The War Experiences of an Australian Infantryman in France, 1916–1919*, ed. Will Davies. Milsons Point, NSW: Random House Australia.

Maddern, Peter. 2021. *The Eye of Wilkins: The Complete Photographic Retrospective of George Hubert Wilkins*. Adelaide: Peacock Publications.

Manning, Frederic. 1999. *Her Privates We*. London: Serpents Tail. First published 1929.

Nelson, Megan Kate. 2005. *Trembling Earth: A Cultural History of the Okefenokee Swamp*. Athens, GA: The University of Georgia Press.

Overy, Richard. 2005. Allied Bombing and the Destruction of German Cities. In *A World At Total War: Global Conflicts and the Politics of Destruction, 1937–1945*, ed. Roger Chickering, Stig Förster, and Bernd Greiner, 277–295.

Washington, DC: German Historical Institute/Cambridge: Cambridge University Press.

Remarque, Erich Maria. 1996. *All Quiet on the Western Front*. Translated by Brian Murdoch. London: Vintage. First published in German in 1929.

Shaara, Jeff. 2004. *To the Last Man: A Novel of the First World War*. New York: Random House.

Sherman, William. 1990. *Memoirs*. New York: Library of America. First published 1886.

Sherry, Vincent, ed. 2005. *The Cambridge Companion to the Literature of the First World War*. Cambridge, UK: Cambridge University Press.

Sloterdijk, Peter. 1987. *Critique of Cynical Reason*. Translated by Michael Eldred. Minneapolis, MN: University of Minnesota Press. First published in German in 1983.

Sloterdijk, Peter. 2009. *Terror from the Air*. Translated by Amy Patton and Steve Corcoran. Los Angeles, CA: Semiotext(e). First published in German in 2002.

Sloterdijk, Peter. 2016a. *Selected Exaggerations: Conversations and Interviews 1993–2012*, ed. Bernhard Klein. Translated by Karen Margolis. Cambridge, UK: Polity. First published in German in 2013.

Sloterdijk, Peter. 2016b. *Spheres, Volume 3: Foams: Plural Spherology*. Translated by Wieland Hoban. Los Angeles, CA: Semiotext(e). First published in German in 2004.

Sloterdijk, Peter. 2018. *What Happened in the 20th Century*. Translated by Christopher Turner. Cambridge, UK: Polity Press. First published in German in 2016.

Taylor, Frederick. 2004. *Dresden: Tuesday, February 13, 1945*. New York: HarperCollins.

Walter, George. 2006. *The Penguin Book of First World War Poetry*. London: Penguin.

Willis, Ann-Marie. 1988. *Picturing Australia: A History of Photography*. Sydney: Angus & Robertson.

Winter, Jay. 2004. Introduction. In *Under Fire*, ed. Henri Barbusse, vi–xix. London: Penguin.

CHAPTER 8

Fire and Air in the Sublime Bombing of World War II Aerial Warfare

Abstract The mixing of the elements of fire and air took place in the sublime bombing of World War II aerial warfare. War in the air was theorised in military texts and science fiction. It was also enacted in the aftermath of aerial bombing in the terror of the ruinous landscape below. WWII was fought over (in two senses) the city: it was fought on and over the ground; and in the air above the city. The theatre of operations was through the countryside to the cities and over the cities. WWI was fought in the country and over the country; it was also fought on, in and sometimes under, the ground. The landscape of trench warfare in WW 1 was a battlefield for land-based military combatants of artillery and infantry pitted against each other in the country, whereas the landscape of fire-bombed cites in WWII was a battlefield for airborne military operations of planes directed against civilians in the city, such as the fire-bombing of German cities, including Hamburg, a city set in a swamp. A first-hand, eye-witness recount of Hamburg emphasised the uncanny. Both world wars had an effect on the four elements and the six senses.

The landscape of trench warfare in World War 1 was a battlefield for land-based military combatants of artillery and infantry pitted against each other in the country, whereas the landscape of fire-bombed cites in World War II was the result of airborne military operations of planes directed against civilians in the city. WWI was fought in the country and over the

R. Giblett, *Living with Nature in the Anthropobscene*,
https://doi.org/10.1007/978-3-032-12822-5_8

country; it was also fought on, in and sometimes under, the ground. The theatre of operations was the countryside with its fields, hills, valleys, villages, woods and wetlands. The immediate strategic objectives were positions in the country: that ridge or salient or village. Paris was a long term and remote German objective. London was bombed in 1915 and 1917, and 1300 people were killed as a result of 225 tons of bombs being dropped on it (Bourke 2005, 224). This tonnage and these casualties pale into insignificance compared to WWII.

8.1 Air War

WWII was fought over (in two senses) the city; it was fought on and over the ground and sometimes in the air. The theatre of operations was through the countryside to the cities and over the cities. Paris and Vienna were immediate German objectives, London and other British cities, targets; later Dresden and Hamburg, and many other German cities were immediate Allied objectives, and targets, as were Tokyo, Hiroshima and Nagasaki. All of the major battles of WWI were land battles; some of the major, if not decisive, battles of WWII were air battles, such as the Battle of Britain fought in the air. Indeed, for Albert Camus (2006, 45) writing in 1944 four years after the Battle of Britain, "Germany lost the war when it lost the Battle of Britain."

Trench warfare was a new kind of warfare whose horrors were largely unimagined before they took place. Aerial bombing was a new kind of warfare whose terrors 'science fiction' writers, such as H. G. Wells, and theorists of air war, such as Giulio Douhet, imagined before they were enacted. Wells's *The War in the Air*, first published in 1908, is a prescient science fiction about the eponymous topic of its title. Bourke (2005, 225, 6) calls it a "prophetic novel" with its "science-fiction horror." Reflecting back some years later in prefaces he wrote for various editions, Wells (2005, 278) noted the precision of his insights that "with the flying machine war alters in its character: it ceases to be an affair of 'fronts' and becomes an affair of 'areas'." Fronts are two dimensional with length and breadth on the ground, whereas areas are three-dimensional spaces with height in the air as well. Flying machines made war possible for Wells (2005, xxix) "in three instead of two dimensions." War was no longer confined to "a restricted area called the Seat of War. Now the whole atmosphere was the Seat of War" (Wells 2005, 174), not only because planes flew in it, but also because it provided one of the necessary components for fire (oxygen). Wherever there is air there is the possibility of fire; and wherever there is air is the possibility of flight and the possibility of planes carrying devices that can cause fire. Aerial warfare creates the terrifying possibility that the air we breathe to sustain life will become the medium of death and the means to destroy life. This terrifying possibility was realised in WWII. Aerial warfare also meant "the disappearance of the theoretical immunity of non-combatants." Flying machines flew over fronts to strafe or bomb targets behind enemy lines. As modern industrial warfare is total warfare fought by a nation at arms whose labour power is mobilised for the war effort, non-combatants become targets.

Besides these strategic insights, Wells's novel also highlighted the fiery destruction and human dimension of aerial warfare with "the smashing impact of bombs" and "the grotesque, swift onset of insatiable conflagrations." From "it all," the central character of the novel, Bert Smallways, "felt detached, disembodied" (Wells 2005, 140). The use of the third dimension in warfare enables the bomber and aerial witness to feel removed and uninvolved in the destruction occurring on the ground below them. It meant that "men […] poured death and destruction upon the homes and crowds below" (149) but felt detached from it.

Wells was well aware of the paradox of "the enormous powers and grotesque limitations of aerial warfare," in particular that "it was impossible to subdue the city except by destroying it" (149). This deadly paradox haunts all aerial warfare, particularly the strategic bombing of cities in WWII. For Wells, "the application of science to warfare" meant that "it was unavoidable that great cities should be destroyed" (149). How unfortunately right he was; how much death and destruction could be laid at the doorstep of the "unavoidable."

Trench warfare was a new kind of warfare whose strategies and tactics were developed on the ground and in practice as the old ones, such as the cavalry charge, were outdated and ineffective. Aerial bombing was a new kind of warfare whose strategies and tactics were devised in the airy heights of military theory before they were enacted. The primary text of aerial warfare and its strategy is Douhet's *Command of the Air* first published in 1921. This book was reprinted by the US Office of Air Force History in 1983, which testifies to its status as a classic of military theory and strategy for the practice of aerial warfare. Douhet (1983, 3) was aware that "aeronautics opened up to men a new field of action, the field of the air" and that this constituted "a new battlefield." A dictionary definition of an aeronaut is of "a traveller through the air," either a balloonist or pilot, and so not a passenger as such. Aeronautics is "the science, art, or practice of controlled flight through the air." The air traveller controls the air vehicle unlike the air passenger. Part of the attraction of aerial warfare for Douhet (52) is that "the air is a uniform element everywhere" subject though, of course, to the vagaries of the weather and so to the speed of air flow and the presence or absence of that other element—water—in the air in the form of rain, hail, humidity, sleet or snow. The plane for Douhet (82) makes "man [] master of space" as "all space is an unlimited roadway for planes" (89). The trope of the road characterises civilian transportation whereas military mastery of space means that it is a largely unlimited theatre of operations for war.

The new battlefield of the field of the air transcended "the surface of the earth" (4 and 7; see also note, 27) and created a new medium for warfare. Artillery shells and incendiary projectiles had been propelled through the air for hundreds of years. With the advent of the manned aeroplane, the projectile could be launched in the air at targets on the ground and other targets in the air. The vertical dimension of the battlefield was increased aeronautically and made the way for their extension astronomically with satellites and 'Star Wars.' Aeronautics enables 'man' to break the ties that tied him to the surface of the earth (7) and to transcend it in an act of sublimation (as we have seen in Chap. 3).

No technological development is without its effects and impacts[1] on its users, and targets, and the use of planes in war is no exception. This holds true not only for pilot and passenger, bomber and bombed, but also for war itself. Prior to the use of planes in warfare there was for Douhet (1983, 9) "a legal distinction made between combatants and non-combatants." With the advent of the use of planes in war "there will be no distinction any longer between soldiers and civilians" (10). Why? Because planes can "bomb our cities" (10), ostensibly in search of military targets, but given the embededness of military facilities and factory supplying the military within cities and the mobilisation of the civilian workforce for the total war effort (and given that targeting will never be 100% accurate and reliable) civilians and civilian facilities will be bombed. "A nation at war" (9) means that all members of the nation are participants in the war, including as targets.

A nation at war in the air (as well as on the sea and the ground) means that the sky is the space in which a threat may be posed to soldier, sailor,

[1] For discussion of the impact metaphor, see Giblett (2008, 23–24).

pilot and civilian alike. All four are linked in the common element of the air. The soldier on the ground, the sailor on the high seas and the civilian in the city or country were previously separated, unless the soldier or sailor invaded the place where the civilian lived. With the advent of planes in war, this separation no longer applied as they, and the pilot, shared a common element. This commonality of the air as an element in, and a medium for, warfare meant that the airplane (and not its weaponry or bomb load) for Douhet (15) in his day is "the offensive weapon par excellence." In our day it is the cruise missile or smart bomb. All are characterised by the use of air as the element through which an offensive weapon of war is conveyed.

As the air poses such a potential threat from one's enemy and as it is the element through which to pose a threat to one's enemy, it becomes a major battlefield, and a stake in battle. The plane in war increases the pressure for the pre-emptive strike. The use of the element of the air in warfare changes the use of the elements of earth, water and fire. The surface of the earth is, for Douhet (1983, 19):

> the coastline of the air. The conditions pertaining to both elements, the air and the sea, are analogous; so that the surface of the earth, both solid and liquid, should be defended from aerial attack, not be scattering guns and planes over its whole extent, but by preventing the enemy from flying. In other words, by 'conquering the command of the air.'

This is Douhet's mantra. For him "to have command of the air means to be in a position to prevent the enemy from flying while retaining the ability to fly oneself" (24). The result of this command is simple, and can be summed up in one word: "to have command of the air is to have *victory*" (25; his emphasis).

Douhet's mantra could be rephrased in the following terms: the best form of defence (of ones cities, factories, civilians) is attack (as he says later (52)) and the best means of attack is the plane. Also for him the best means of defence of "the air arm" is attack (42). He finally concludes that "the basic principle which must govern the development of aerial warfare "is that "aerial warfare admits of no defence, only offence" (55). The air becomes a liminal zone of attack alone unlike the coastline, which is both a zone of attack and defence. Through the air zone an attack may come and defence must be made preferably (in military terms) by attacking the enemy first. This is the doctrine of the pre-emptive strike that reaches its nadir in MAD, 'mutually assured destruction.'

The targets for such a pre-emptive attack could include what Douhet (20) calls "the civilian population of congested cities" with the attacker "making no distinction between military and non-military objectives." The result for him is inevitably that "the decisive blows will be directed at civilians" (61). From the air the problem for bombers is to distinguish military from non-military targets, and military from non-military personnel, especially when some civilians have been mobilised in total war and the nation is at war. The bombs for attacking such targets for Douhet are of three types: explosive, incendiary and poison. With the last type outlawed in the aftermath of WWI, the former two were used in WWII with destructive and deadly force. For Douhet (20), "the explosives will demolish the target, the incendiaries will set fire to it," and destroy it. "Demolish" implies built structures, but as civilians are also targets, so human beings, and bodies, are "demolished" by explosives, and set fire to and destroyed by incendiaries. With the use of incendiary bombs in air warfare, the element of fire is introduced into the equation with the element of air. Or perhaps more precisely, incendiary bombs *re*introduce fire into warfare with devastating and destructive impact. Earth, air, fire and water, the four elements, are theatres of warfare (in the case of earth, air and water), and even weapons in warfare (in the case of fire).

Command of the air follows, and is made possible by what Douhet (29) calls "man's [*sic*] conquest of space." Command of fire is made possible by 'man's' conquest of matter. The former took place through the science of aeronautics, the latter through the science of chemistry. This conquest took place not only in the fire of the internal combustion engine, but also in the fire of incendiaries. Fire was both contained in a small explosion within an engine and spread in a large immolation with an incendiary bomb. Fire was both harnessed and unleashed. "What determines victory in aerial warfare "for Douhet (44) is "fire power," the power of fire delivered in explosive and incendiary bombs as quickly and accurately as possible.

War in the air not only reshapes the element of air and has impacts on the earth below and what is on it (such as cities) and on those who live on it (and in them). It also reshapes the relationship between 'men' and machine. The warplane (fighter or bomber) is like an airborne tank in that the operator is cocooned in a metal box. Both are indicative of the way in which modern industrial world warfare is fought not only by "masses of men and machines" as Douhet (44 and 84) puts it twice, in tank or plane squadrons and formations (45), but also men *in* their machines. Both the warplane and the tank have armor plating for protecting "the vital parts" (45) of the machine, and the men inside them. 'Men' and metallic machine form a mass, a unitary being with 'vital parts,' flesh and metal. Warplane and war fighter come together for Douhet (69) with the "forging of an arm" of the monstrous military man-machine. They also come together with the introduction of two-way radios in WWII to form the cyborg, the man-machine-communication technology composite. The forging of "air arm" of the military, the development of an 'airforce' to deploy with the land forces of the artillery and infantry and the sea forces of the navy, is made possible by the forging of the individual warplane. This machine, for Douhet (77), "man has forged out of his genius." It is thus a Bachelor Machine from a Bachelor Birth. It is a product of cerebral parthogenesis which cannot create life—only destroy it as occurred in WWII.

8.2 Firebombing

Writing on the firebombing of German cities by Allied bombers, W. G. Sebald (2003, 19) argues "the war in the air was war pure and undisguised." In this pure air war "the innermost principle of every war, which is to aim for a wholesale an annihilation of the enemy with his dwellings, his history, and his natural environment as can possible be achieved," was pursued. This principle was exercised by war being taken to the new heights achievable by fighter and bomber planes which not only produced "a kind of mobile front line" as Sebald (2003, 17) puts it to and fro across the horizontal dimension of space, but also up and down within the vertical dimension. Planes and balloons had been used in WWI but it was predominantly an earth-bound war that took place on, and in, the earth, in its depths beneath the surface as we have seen.

WWII was less earth-bound and took place on, and above, the surface of the earth. The results were different too with whole French towns and villages obliterated by artillery bombardment and their residents killed or displaced as refugees in WWI, whereas in WWII many German cities were ruined from firebombing and their residents burnt to death or displaced too. Both landscapes, however, were described as 'lunar landscapes' as we have seen with WWI and as Sebald remarks with WWII. The lunar landscapes of both WWI and II were both produced by bombardment and both were wastelands (Sebald 2003, 74) and products of modern industrial warfare (64), or more precisely "the machinery of annihilation operating on an industrial scale" (96).

Yet they there were different in that the lunar landscape of WWI was characterised by craters full of water on and below the surface of the earth whereas the lunar landscape of WWII was made up of "mounds of rubble" (Sebald 2003, 36, 74) on and above the earth's surface. The former was wet and cold and the latter dry and hot. The countryside of WWI trench warfare was a wastewetlandscape of artificial 'swamps' whereas the bombed cities of WWII aerial warfare were "a landscape of ruins" (46). Although Blunden (2000), Hurley (1917) and Thomas (1978) comment repeatedly on the ruins of WWI, these were of individual buildings and villages, not of large proportions of cities as in WWII. Sebald called some areas of Germany during WWII "that vast wilderness, perhaps the most horrifying expanse of ruins in the whole of Europe" (2003, 30). From France as 'swamp' to Germany as ruin marks a profound, elemental shift in the exercise of armament, in the deployment of fire-power and in the experience of war.

It also marks a shift in the type of landscape produced from feral swamp to ruined cities, from human-made countryside remade into human-made swamp to human-made city remade into human–made ruin. A ruin is defined by Solnit (2005, 88) as "a human construction abandoned to nature." A bombed city is an artificial ruin, a human construction destroyed by culture. A feral swamp is a native construction abandoned by humans. Both swamp and ruin stand as figures for the unconscious. Just as the swamp is a figure for the cultural unconscious (see Giblett 1996), so ruins for Solnit (2005, 89) "become the unconscious of the city." Out of both new life can spring as Solnit (2005, 90) argues that "with ruin a city comes to death, but a generative death like the corpse that feeds flowers," just as the corpses did in the battlefields of WWI.

Cities and citizens, generally well-away from troops fighting on the ground, became the front line in aerial warfare. In modern war, for Friedrich (2006, 357), "the difference between soldier and civilian no longer existed; everyone was a warrior." In the air war everyone was a combatant and a target. "Civilian warriors," as Friedrich (2006, 358) calls them, "were defenceless; they had no weapons." The front line was not merely mobile in time and space, but also socially mobile across the distinction between soldier and civilian. Civilians for John Pimlott (1988, 123) were "the new front line in total war" with devastating consequences for them and their cities and homes in them.

Germany, for Camus (2006, 40) writing during WWII, was "resigned to the appalling destruction that is raining down on their cities." Camus later conjured up "an apocalyptic image" of "its cities [that had been] transformed into shapeless rubble" (229). What had rained down from the sky had destroyed cities on the ground. Camus visited occupied Germany shortly after the Armistice and what he saw "reinforced the premonition" as he found "the rubble and the barren fields ravaged by war and dotted with military cemeteries overshadowed by ungenerous skies." He concluded that "this indeed is the land of the dead." Resigned or not to the destruction, the fact remains that the sky above was no longer just the generous source of life-giving rain, but also the greedy medium of death-dealing destruction on cities and citizens alike in the anthropobscenity of modern industrial warfare.

The deployment of what Virilio (2005, 15) calls "our aero-naval modernity" against cities entailed what he also called "the 'aeropolitics' of a mass extermination of cities" (Virilio 2007, 8). This mass extermination of cities destroyed buildings and their inhabitants, a kind of genocide of cities, a citi-cide, or what Helphand (2006, 246) calls "urbacide," "the wilful destruction of cities." Just as WWI turned the country into a 'desert,' WWII turned the city into a 'desert.' For Virilio (2007, 7) "a human environment [was] turned into a desert through the annihilation brought about by air raids." Like WWI, the 'desert' of WWII was not a living 'natural' ecosystem, but a human-made wasteland. Whereas the 'desert' of WWI was wet, muddy and cold, the 'desert' of WWII was dry, rubbly and hot. This mass extermination of cities occurred at the tail end of WWII. For Virilio (2005, 15), "after Dresden, and especially after Hiroshima and Nagasaki, this 'aeropolitics' turned into a cosmopolitics of nuclear terror in the Cold War with the *Anti-City* strategy."

What Sloterdijk (with Heinrichs 2011, 117) called "the lightning blitzes of Hiroshima and Nagasaki" superseded and upped the ante of the firing blitzes of European cities in WWII from the incendiary to the thermonuclear. The nuclear bombing of Hiroshima and Nagasaki heralded for Bourke (2005, 363) the inception of "the Age of Terrorism," which then thrived during the Cold War. The "terror from the air" (Sloterdijk 2009, 2016, 85–183) of fire-bombing led to the age of terrorism of the threat of thermonuclear annihilation in a further appropriation, escalating and upscaling of the sublime Law of the Father.

This age with its anti-city strategy, however, was by no means pro-country, but merely replaced and extended the anti-country strategy that had culminated in the landscape of trench warfare to the anti-city strategy that culminated in the landscape of ruined cites. Our aero-naval hypermodernity is both anti-city and anti-country. The anti-city strategy of WWII was carried out by the fire-storming of strategic bombing to create dry, ruinous wastelands, whereas the anti-country strategy of WWI was carried out by the earth-moving of artillery bombardment and trench construction to create the wetwastelands of artificial wetlands.

The ruination of European cities began in 1940 with the German Blitz of Coventry and London, increased with the Allied bombing of Hamburg in 1943 (Lowe 2007), and culminated in the destruction of such cities as Dresden (Taylor 2004) and Würzburg (Knell 2003) in 1945. Goebbels evoked the bombing of London in 1940 in Dantesque terms as "an inferno of unimaginable extent. The city is coming to resemble a hell" (cited by Taylor 2004, 100). Yet this bombing was limited and the resulting fire was half a square mile in area. It was officially named a "conflagration." For Taylor (2004, 108) it "was the nearest London came to what later would be called a firestorm." For Grayling (2006, 18) "the first ever firestorm caused by bombing" was in Hamburg in July 1943, not in London in 1941.

Yet a two-part BBC-TV documentary on the Blitz claims that London more than came close to a fire-storm as it is subtitled *London's Firestorm.* Some of the interviewees describe the event as a firestorm. 'Bomber' Harris, one of the architects of the allied firestorm, witnessed the London Blitz of December 1940 and remarked of the Germans that "they are sowing the wind" (cited by Taylor 2004, 109; Grayling 2006, 49). He assumed that his biblically literate audience would supply the second half and draw the conclusion. The rest of this Biblical verse (Hosea 8:7) goes on to prophesy that those who sow the wind would reap the whirlwind—as indeed the Germans did as one witness remarked of the firebombing of Dresden and the resulting firestorm (Taylor 2004, 267). Yet as Friedrich (2006, unpaginated 'Afterword for American and British Readers') points out, "a characteristic of air wars is that those who sow the wind do not reap the whirlwind and those who reap the whirlwind did not sow the wind." Indeed. The rest is, as they say, history, including the history of human pain and suffering, as well as the history of theology and the theology of history not commented upon by theologians.

8.3 Sodom and Gomorrah

The book of Hosea for J. Gordon McConville (2012, 344) "evinces a kind of theology of history," which goes back to the history of theology too. Rather than in the terms of Hosea 8:7, aerial bombing was seen from its beginnings in the early twentieth century in terms of the biblical story from the book of Genesis about the destruction of the sinful cities of Sodom and Gomorrah. Beginning with the first aerial bombing by Italians of Turkish troops in 1911 in North Africa, "across Europe the popular imagination," according to Lowe (2007, 42), was "fired with images of a man-made Sodom and Gomorrah." This differed from the divine destruction of the two cities in the biblical book of Genesis, but how it differed is not clear except in terms of the agent of destruction. What was certain is that 'man' had become God by taking on a divine function and power of destruction (just as earlier 'he' had taken on the divine function and power of creation in developing cities).

The story of the destruction of Sodom and Gomorrah has for J. A. Loader (1990, 38 and 43) "an anti-urban tendency" and "bears evidence of an anti-urban attitude" in which "the sinfulness of life in the city is contrasted with the rustic virtue of hospitality found in nomadic society." Lot, "showing as he does respect of the nomadic ideals, is presented as a tragic illustration that the nomad way of life is not tolerated in the city." Lot's wife, showing as she does disrespect for God's interdiction not to look back at the destruction of the cities, is presented as a tragic illustration that God does not tolerate disobedience. God wins (and Lot and the nomadic way of life win too); Lot's wife loses (and sinful cities lose too). God uses the power of nature, such as seismic activity and chemical reaction, to destroy the cities. The cities were overturned, which for Loader (1990, 41) "points to an earthquake," and "the sulphurous rain [that descended on them points] to volcanic activity."

The images of the divine destruction of Sodom and Gomorrah by "fire and brimstone" (Genesis 19: 24 (King James/Authorised Version)) were pressed into the service of the potential bombing and destruction of British cities in the preamble to WWII. By the 1930s, as Lowe (2007, 47) puts it, "the most enthusiastic champion of bombing" in England was, hardly surprisingly, "forced to admit to a widespread pessimism among his contemporaries, who foresaw "a fate comparable to that of Sodom and Gomorrah" for British cities." The fate of Sodom and Gomorrah was mainly to befall German cities as pre-emptive strikes designed to avoid the repetition of the "man-made" destruction of Sodom and Gomorrah with

British cities. Yet the results for Lowe (2007, 48) were "no longer a vision of Sodom and Gomorrah," but "a vision of Armageddon," the battle culminating in and at the end of the world as related in the book of Revelation, the final book of the New Testament.

The bombing of Hamburg in 1943 was called "Operation Gomorrah," as noted by Martin Harries (2007, 84) and Sebald (2003, 26). Sebald (2003, 57) later uses "Operation Gomorrah" as a shorthand, non-euphemistic term for the bombing of Hamburg with all its attendant horrors and terrors. He also later invokes what Walter Benjamin (2003, 392) calls "the Angel of History" looking at piles of wreckage and debris not unlike that produced by "Operation Gomorrah" against Hamburg (Sebald 2003, 67–68; Sloterdijk 2016, 127). By contrast with Lot's wife who looks back (to the past, to a destroyed place, to a lost time and place) and is destroyed for doing so, the Angel of History for Benjamin (based on Paul Klee's picture *Angelus Novus*) in his essay 'On the Concept of History' looks back into the past as "his face is turned toward the past," and the destruction that took place in it, with hir back turned towards the future and with the winds of progress beating into hir face and pushing hir into the future. Rather than "a chain of events that appears before us," "the Angel of History" for Benjamin (2003, 392) "sees one single catastrophe which keeps piling wreckage upon wreckage and hurls it at his feet [...] This storm drives him irresistibly into the future to which his back is turned, while the pile of debris before him grows toward the sky." Rather than a series of catastrophes proceeding chronologically one after the other through history, Benjamin's "Angel of History" sees one single catastrophe of wreckage in time.

The Angel of History looks back into the past as s/he moves backwards into the future. S/he does not consider the future to be a prospect—pleasing or terrifying. S/he does not consider the future at all because the future is inconsiderable. The future is behind hir. The past is before hir. The Angel looks back at, and into, the past as a "pass-pect." The Angel even looks beyond the past, past the past, as Samuel Weber (2008, 226; his emphasis) has suggested:

> the glance of Klee's *Angelus Novus* may be directed in the general direction of the viewer, but [...] his eyes are riveted *elsewhere:* off to one side [...] The New Angel glances at us and yet past us [...] In thus looking past the past, such a glance leaves room for a different kind of future, one that would not lead "back" to the place from where we have come.

The Angel does not look at us as that would fix us in its gaze at a particular moment in time whereas we are always players in time suspended between a past we cannot change, a present we cannot escape from, and a future we cannot know.

The single catastrophe of wreckage upon wreckage occurs not only in time, in history, but also in space, in a place, in geography (as the biblical story shows). The Angel of Geography sees one single catastrophe of wrecked and lost places, such as wetlands dredged, filled, impounded, or embanked, and so 'reclaimed' (from what? for whom?), cities and suburbs set in them and being re-reclaimed by them in storms and floods (and increasingly so in the age of global boiling), and cities, towns, and villages ruined in watery floods, or in fiery destructive cataclysms like Sodom and Gomorrah and their reprise in the aerial bombing and firestorms of WWII, history repeating itself with a difference.

As well as the Angels of History and Geography—in one case whose divine and eternal point of view of wreckage is unattainable by mere mortals and in the other whose omniscient and omnipresent point of view of wrecked places is also unattainable by mere mortals —the Angel of Corporeality lives bodily present in mind and place, a non-dualistic, holistic way of being probably unachievable by modern urban human beings. While the Angel of History turns "the threatening future" into "a fulfilled now" as Benjamin (1979, 99; 1996, 483) puts it in *One-Way Street*, the Angel of Geography turns threatening places into a fulfilled here and the Angel of Corporeality turns the threatening mind and/or body into a fulfilled, whole mind-body. From this point and place in space-time, these Angels look over disastrous places and ruined bodies and minds, as well as look back at the disastrous past with mixed fear of, and nostalgia, for the past and lost places. They are not afraid and they are in no danger of being turned into a pillar of salt, like Lot's wife, for they are not fixated on the past. They face the past, present, places, and the future with fear and hope in embodied minds and mindful bodies.

Humans facing the past, present, places, and the future can do so with hope and fear like the Angels, our better selves. One human facing the future in fear and despair before WWII saw it in the terms of Sodom and Gomorrah as some who experienced the aerial bombing and firestorms of the war. Sloterdijk (2009, 103; 2016, 176) cites how the pre-war anti-establishment Viennese journalist Karl Kraus in 1936 in the last issue of *Die Fackel* published four months before his death gave up his fight against what he called "the air of Sodom." Kraus sensed that the oxygen had been sucked figuratively, culturally and politically out of the life-giving air of pre-war Europe into a fiery conflagration of destruction several years before this occurred physically in fact. Perhaps he also prophetically or presciently foresaw aerial fire-bombing in WWII that sucked oxygen out of the air literally. On a post-war tour of Europe, Stephen Spender (cited by Harries 2007, 132, n.24) commented that "we can regard Germany as Sodom." For Kraus Germany was already was Sodom before the war, and not just because it was sinful. The pre-war atmosphere of Europe was already toxic; bombing only enacted it and brought it to fruition. Sloterdijk (2012, 25) relates elsewhere that Edmund Husserl, also in the 1930s, "had already witnessed the beginning of the European spirit's descent into hell." It was descending into hell again after it had only recently emerged from the hell of WWI and the trench landscape of German idealism. It was descending into the hell of WWII and the bombarded landscape of German fascism and its allied counter-force. Both were anthropobscenic.

All those writers who invoke Sodom or Gomorrah do not mention Lot's wife as if they were afraid of being turned into pillars of salt themselves for looking back at the destruction of German cities or European culture.

Some historians of the past (as distinct from historians of the present) seem to be unable to take their own advice, or are at least unable to be reflexive about their own position and role looking back into the past. As the epigraph to the chapter on survival in his book about the bombing of Hamburg, Lowe (2007, 277) cites the warning of Genesis 19:17: "Look not behind thee [...] lest thou be consumed." He does not follow this advice, nor heed this warning himself. Perhaps he does not think it applies to him and only to the survivors of the bombing of Hamburg. He explains in an endnote at this point that these are "the words of the angels to Lot, as they saved him from the destruction of Sodom and Gomorrah. Lot's wife did look back and for her sin was turned into a pillar of salt," as Genesis 19: 26 goes on to relate (it is noteworthy that Lowe (2007, 403, n.1) relegates Sodom and Lot's wife to a note here and that Harries (2007, 132, n.24) also relegates "Germany as Sodom" to a note).

Looking back has both a spatial and temporal dimension and inflection in retrospection. Just as prospect has both a spatial and temporal dimension of looking over a place or forward into time, so is retrospect as it involves looking back into the past and at a lost place; just as prospects can be either pleasing or displeasing spatially and temporally, so can retrospects. Lot's wife not only looks back at the place that has been destroyed, but also tries to look back into the past that has also been destroyed. For this double sin she was turned into a pillar of salt.

8.4 The End

Rather than the historian, the first-hand, eye-witness "forthteller" gives the most compelling and harrowing account of the whirlwind when Hans Erich Nossack (2004) speaks truth to power and tells his own story of the firebombing of Hamburg. Nossack (2004, 1–2) witnessed the destruction of Hamburg as a spectator from outside the city. Hamburg was "the first big city to be annihilated" in WWII according to Nossack (2004, 32) who witnessed its destruction as a spectator from outside the city. He bore witness to the destruction in his book called *The End*. The title refers not only to the end of the city as he knew it, but also the end of the world in apocalyptic tones. What he calls "the ruin of Hamburg" (6) is also variously described in apocalyptic terms as "the abyss" (6), "the end" (8) and "the netherworld" (11). Not merely "anti-city," its destruction also pitted nature against nature for "even nature had risen up in hatred against herself" (11). Aero-naval modernity is anti-nature where two forces of nature (fire, storm) are both harnessed by the militaristic men and unleashed against nature. The result for Nossack is "the raging of the world against itself" and "the earth writhed in agony" (15). These are the death throes of "raging destruction" (13) inflicted by the world and the earth upon itself by a world- and earth-hating military whose end (in 2 senses) is "the ruined world" (13). "The horror" of "the disaster" (16) was that it was both natural and cultural, the forces of nature directed against both nature and culture by natural and cultural human beings against other natural and cultural human beings.

Part of the horror for Nossack was the fact that the refugees fleeing from the city "brought with them an uncanny silence. No one dared question these mute figures" (17). They had ceased to be human, to be natural and cultural beings, because they were deprived of speech and had become speechless automatons who mimicked and mocked human behaviour in uncanny fashion. The fact that this disastrous destruction of Hamburg was "the end" for its residents of 1943 meant "we no longer have a past" (23), only a present and future rent from any connection with the past. Perhaps they did not even have a present and future as "we no longer had any time at all, we were outside of time." In other words, they transcended time into eternity (40). Space and time ceased to exist as they had been known.

Not only was the spatial landscape transformed, but also the temporal one as well: "the infinite behind man wafted unhindered in the endlessness before him and hallowed his countenance for the passage of what is beyond time" (29). Space and time are collapsed together into the landscape of eternity and infinity of the landscape of bombed cities (as we saw with landscape of trench warfare in the previous chapter). The less, or more, than human figures, who people this landscape are not only mute automatons, but also icons with eyes "grown larger and transparent" (29) through witnessing the horrors of fiery destruction of their city and the (its, their) past.

Hamburg has a fiery and watery history as its centre has been destroyed by fire four times and, as Lowe (2007, 3, 5) puts it, "it is situated on a fluvial plain, most of the ground is little better than a marsh, and it is prone to flooding." Hamburg is like a number of other cities built on wetlands, such as London, Paris, Venice, Berlin, St Petersburg, New York, New Orleans, Washington, Chicago, Toronto, Jakarta, and Perth, Western Australia (see Giblett 1996, 2016). Some of these cities drained and filled their wetlands, whereas others, like Hamburg, built canals and waterways so, as Lowe (2007, 3) goes on to argue, "the element that most dominates the city is water." The domination of water made the city ironically more susceptible to fire-bombing as Lowe (2007, 32) points out:

> the soggy, waterlogged soil upon which Hamburg was built meant that whole districts were devoid of cellars – the high-water table of the Elbe floodplain would simply have swamped them.

The very element of water that quenches fire made the city, or at least its citizens, more vulnerable to fire, or at least to firebombing from the air. Of the Hamburg firestorm, Lowe (2007, 182) says that it was "one of the biggest man-made fires the world has ever seen." Later he says that "no other large fire in recorded history has ever equalled it in intensity" and that it was "the greatest firestorm the world has ever seen" (Lowe 2007, 185).

8.5 The Elements and the Senses in World Warfare

The transformation of the land in both wars had its effect on those who fought these wars and were affected by them. That reshaping took place at the fundamental level of the arrangements between the four elements of earth, air, fire and water as we have seen in this and the previous chapter. The reconfiguration of the elements in world warfare had implications for, and effects on, human sensory perception in the feral quaking zones of trench and aerial warfare. Warfare is not only an assault on the human body in general, but also on the senses in particular. The nature of this assault changed in the transition from the trench warfare of WWI to aerial warfare in WWII. Soldiers for Helphand (2006, 33) in WWI "saw, heard, smelled, touched, and even tasted daily, hourly, what in a sane world would only ever be [a] fleeting assault on the senses by the horrific." Bombers in WWII saw, heard and may have smelled, but certainly did not touch and taste, daily what in the insane world of modern industrial warfare was a fleeting and limited assault on the senses by the horrific. The sensory experience of the feral quaking zone of warfare differs also from that of the native quaking zone of wetlands. The swamp can look ugly, smell bad, feel slimy, taste muddy and sound oozy, and produce horror and the uncanny, but its address to the senses can be pleasant, and it is certainly not as horrific nor as terrifying as the assault on the senses that warfare entails. Considering all five senses in relation to the feral quaking zones of industrial world warfare leads to a new embodied environmental aesthetics.

For Giacomo Leopardi, the nineteenth-century Italian poet and essayist, and Italy's second national poet after Dante, the element of air is linked to hearing. He remarked that air is "the element destined for sound" (Leopardi 1982, 357). Air is the element in which sound addresses the sense of hearing and the organ of the ears. Similarly for Immanuel Kant (2006, 47; his emphasis):

> The sense of hearing is one of the senses of merely *mediate* perception. Through and by means of the air that surrounds us a distant object to a large extent is cognised [...] The shape of the object is not given through hearing.

An explosion heard in the distance is largely understood, but the shape of the explosion, its size, extent and impact, is not ascertained through hearing, only through sight.

Air war is a war of sounds. It percusses the sense of hearing and the organs of the ears with the sounds of deafening artillery, humming propellers, whistling and crashing bombs, howling and wailing sirens, punching blast waves, quaking buildings, screams of terror, and raging fire-storms. For Camus (2006, 61) writing during WWII, "the sound of explosions and destruction [is] the very voice of death."

The sounds of war give rise to terror. This association between the noise of war and the sensation of terror is a feature of any war utilising artillery. Goethe (1987, 652) in his account of the Prussian campaign in France in 1792 made the association between the two in the following terms:

> It is [...] remarkable how whatever causes such terrible anxiety [under bombardment] is conveyed to us solely through our sense of hearing: for it is the thunder of the cannon, and the howling, whistling and blasting of cannon balls through the air that is actually the cause of these sensations.

Air war takes the link between bombardment and terror to new heights, to greater extent and for longer periods.

Extending Leopardi's association of air with sound and hearing to the other elements and senses, earth is the element destined for seeing. Earth is the element directed to the sense of sight and the organ of the eyes. Earth war is a war of sights. It assaults the sense of sight and the organs of the eyes with the sights of the battlefield of trench warfare, of wounded or dead soldiers, of sudden flashes, blasted landscapes, ruined buildings, burnt and dismembered civilians, of living lands and cities transformed into piles of rubble and wastelands, of land no longer life-giving but death-dealing, changing from womb to tomb with people trapped or killed in bunkers and cellars. During WWII in the bombed cities of Germany "living space became a death zone" (Friedrich 2006, 84) after it had passed through a quaking zone of shaking and falling buildings.

For Leopardi, "eyesight and hearing, according to the natural order concerning living creatures, are the principal senses." They are the two senses of aesthetics for Hegel (2004, 43), with the eye, as Bloch (1998, 1) argues, "more commanding than the ear." The eye sees and commands from afar. It has a longer range of perception than the ear. The sense of sight for Kant (2006, 48) is "also a sense of *mediate* sensation" like hearing and it is the noblest of the senses for him because it "not only has the widest sphere of perception in space but its organ feels least affected." The eye commands unfeelingly from afar. Eyesight and hearing are also for Leopardi (1982, 367) "the two senses most characteristic of the living." Air and earth are also two elements for the living.

Yet in war, especially in modern industrial warfare, air and earth are elements of the dead and dying. They are elements of an anti-aesthetics of ugly sights and cacophonous sounds. Fire-storms sucked oxygen out of the air until it ceased to be life-giving and became death-dealing. In the words of one eyewitness of the Hamburg firestorm, "the air was actually filled with fire" (cited by Grayling 2006, 83). The bombers produced the sound of "the doom-laden drone in the heavens" (cited by Lowe 2007, 161). The firestorm produced the sounds of "a shrill howling" that "grew into a hurricane" and "a sea of flames" (cited by Grayling 2006, 83). Lowe (2007, 195) notes that "*Flammenmeer*, 'sea of flames,' comes up again and again in accounts of the firestorm" and he cites many instances of this usage. For Friedrich (2006, 84 and 167):

> through heat, radiation, and toxic gases, the very air was being transformed into something unliveable [...] the air that life needs to thrive had been exchanged for something else. The firestorm simulated the atmosphere of another planet, one incompatible with life.

Earth had ceased to be earth, and air, air.

Extending Leopardi further, fire is the element destined for touch, the sense that for Bloch (1998, 157) "gives the strongest impression of reality." The sense of touch for Kant (2006, 46 and 48; his emphasis) is "the only one of *immediate* external perception," but it is also for him "the most limited condition of perception" because of its limited spatial range compared to the sense of sight. Touch gives a strong and deep impression and an immediate perception, but it is limited and narrow in range and scope.

Fire is destined for touch because it radiates heat to the sense of touch and the organ of the skin. Fire war is a war of radiation to the surface of the body and combustion of surface of the body and the earth with burning flesh, buildings, and trees. The skin in fire war, Friedrich (2006, 440) says, "felt the temperature and the blast of air, the build-up of the blaze and the wind that carried it." Fire war also produced toxic gases. The flames, Friedrich (2006, 330) says, "poisoned the air. Combustion generates heat and gases, the main agents that attack the body." And not just the outside surfaces but also the inside surfaces. The nose, Friedrich (2006, 440) says, "registered the fire and odorous gases." Toxic gases generated by burning were the main cause of death (for example, in Hamburg 70 per cent to 80 per cent (Friedrich 2006, 167)).

Water is the element destined for tasting. It drenches the sense of taste and the organ of the palate (mouth). Water war is a war of drenching of the body with torrential rain, of swallowing it and drowning in it, or in mud, water mixed with earth. In fire war, Friedrich (2006, 367) points out, "water had a hard time keeping up."

The four elements of earth, air, water and fire address the four senses of sight, hearing, taste and touch. The fifth sense, the sense of smell, is affronted by all four elements. The four elements impunge, to coin a new word, on the sense of smell and the organ of the nasal mucous membrane (nose). Smell, for Kant (2006, 51), is:

> taste at a distance [...] the pleasure coming from the sense of smell is always fleeting and transient.—But as a negative condition of well-being, this sense is not unimportant, in order not to breathe in bad air [...] the stench of swamps and animal carcasses), or also not to need rotten things for nourishment.

The sense of smell operates for Kant both positively as a sense of pleasure "of the most intimate taking into ourselves" (49) and negatively as a sense of avoidance of taking bad smells into ourselves. A case in point for Kant (50) is filth which, for him:

> Seems to arouse nausea not so much through what is repugnant to the eyes and tongue as through the stench that we presume it has. For taking something in through smell (in the lungs) is even more intimate than taking in through the absorptive vessels of the mouth or throat.

Air can smell bad, such as the smell of bad air or poisonous gas, such as in WWI gas warfare, the smell of chlorine or mustard gas, or "the stench of rotten flesh [...] over everything" so "you could smell the front line miles before you could see it" (Fussell 2000, 49) (and "you could smell the battlefield miles before you reached it" (Helphand 2006, 33)). Fire can smell bad such as the smell of burning organic matter. In a fire war, such as WWII firebombing, the smells of gases and of burning or burnt and decomposing flesh smelt bad (as in the case of eyewitnesses, or nose witnesses (cited by Lowe 2007, 179, 268)). Earth can smell bad, such as the smell of rotting or decaying or decomposing solid organic matter. In an earth war, such as WWI trench warfare, the smell of dead bodies and shit smelt bad. Water can smell bad such as the smell of stagnant water. In water war, such as WWI trench warfare, the smell of water steeped in dead bodies in flooded tranches, the smell of bodily solids, such as shit, and of bodily fluids, such as urine and blood.

Sight and sound give rise to terror and the sublime; smell and touch give rise to horror and the uncanny. For Kant (2006, 49), touch, sight and hearing are "the three highest senses." They are "senses of *perception* (of the surface)," or of "vital sensation" (45), and are "more objective than subjective" (46). For Kant "these three outer senses lead the subject through reflection to cognition of the object as a thing outside ourselves" (48). For him smell and touch are "senses of *pleasure* (of the most intimate taking into ourselves)" (49), or of "organic sensation" (45). They "penetrate the body" (46) and are "more subjective than objective" (46). These inner two senses lead the subject through penetration to the sensation of the object as a thing inside ourselves. Taking a thing inside ourselves can give rise to revulsion or propulsion to get rid of it as soon as possible. Nausea for Kant is "such a strong vital sensation" as the "intimate taking in can be dangerous for the animal" (49). Taste, Sloterdijk (1989, 64) says psychology says, is "the most intimate, the most universal sense of perception." Smell and taste are senses of the depths of the body. The smells of the earth or water, in the air, or of burning and the taste of water or food penetrate into the depths of the body. Sight, touch and hearing are senses of the surfaces of the body. The sight of the earth, the sounds through the air, and the heat or warmth of fire address the surfaces of the body.

All four elements are "destined" for one of the five senses and for one of five organs of the human body (Table 8.1):

The four elements are produced by the mixing of qualities of hot, cold, wet and dry. The four elements give rise to four 'humors,' or human characteristics, or psychological states, in the ancient and Elizabethan schema of things and beings (as we saw in Chap. 3 of the present volume; Table 3.1).

WWII air-fire war mixed these elements in the sublime fire-bombing of cities and produced a hot, dry and gaseous war. It was also characterised by heated exchanges and angry outbursts over the morality of strategic bombing and by a bloody-minded and choleric leader, such as Churchill,

Table 8.1 Elements, senses, and body

Element	*Destined for*	*Sense*	*Body part/Organ*
Air/Gas	Sound	Hearing	Ears
Earth/Solid	Seeing	Sight	Eyes
Fire/Heat	Feeling	Touch	Skin
Water/Liquid	Tasting	Taste	Mouth
All four	Smelling	Smell	Nose

pitted against Hitler, "that vulgar angel of death," as Sloterdijk (2011, 188) calls him. WWI water-earth warfare mixed these two elements in the slimy trenches of man-made swamps and produced a cold war in the literal sense of being alternatively wet and dry depending on the weather, the battlefield alternatively solid or liquid and every stage in-between. It was also characterised by phlegmatic and melancholic soldiers who invoked for Fussell (2000, 139) "the Slough of Despond" as "the only adequate designation for churned-up mud morasses pummelled by icy rain and heavy shells." For him it would be impossible to count the number of times this imaginary place of melancholy and depression associated with native wetlands was invoked by and to describe sullen soldiers stuck in slimy mud (as we saw in Chap. 3). It would be nearly equally impossible to count the number of times "British phlegm" is invoked in Fussell's study of the literature of WWI as the only adequate designation for the unflappable and ironical turns of phrase employed in the literature of trench warfare. The elements and humors are relics of a bygone age, yet they live on as heuristic devices, a conclusion recently echoed by Noga Arikha (2008, 304) for whom "the old humors are gone, but they are still useful, suggestive, and malleable images" used even by recent writers to explain the inexplicable, to figure the horror and terror of modern, industrial world warfare in the Anthropobscene.

References

Arikha, Noga. 2008. *Passions and Tempers: A History of the Humors.* New York: HarperCollins.

Benjamin, Walter. 1979. *One-Way Street and Other Writings.* Translated by Edmund Jephcott and Kingsley Shorter. London: NLB.

Benjamin, Walter. 1996. *Selected Writings: Volume 1, 1913-1926.* Edited by Marcus Bullock and Michael Jennings. Cambridge, MA: Belknap Press of Harvard University Press.

Benjamin, Walter. 2003. *Selected Writings: Volume 4, 1938-1940.* Edited by Howard Eiland and Michael Jennings. Cambridge, MA: Belknap Press of Harvard University Press.

Bloch, Ernst. 1998. *Literary Essays.* Translated by Andrew Joron and Others. Stanford, CA: Stanford University Press.

Blunden, Edmund. 2000. *Undertones of War.* London: Penguin. First published 1928.

Bourke, Joanna. 2005. *Fear: A Cultural History.* London: Virago.

Camus, Albert. 2006. *At Combat: Writings 1944-1947.* Edited by Jacqueline Lévi-Valensi and Translated by Arthur Goldhammer. Princeton, NJ: Princeton University Press.

Douhet, Giulio. 1983. *Command of the Air.* Translated by Dino Ferrari. Washington, DC: Office of [US] Air Force History. First published in Italian in 1921.

Friedrich, Jorg. 2006. *The Fire: The Bombing of Germany 1940-1945.* Translated by Alison Brown. New York: Columbia University Press.

Fussell, Paul. 2000. *The Great War and Modern Memory.* New York: Oxford University Press. First published 1975.

Giblett, Rod. 1996. *Postmodern Wetlands: Culture, History, Ecology.* Edinburgh: Edinburgh University Press.

Giblett, Rod. 2008. *Sublime Communication Technologies.* Basingstoke: Palgrave Macmillan.

Giblett, Rod. 2016. *Cities and Wetlands: The Return of the Repressed in Nature and Culture.* London: Bloomsbury.

Goethe, J. W. von. 1987. Campaign in France 1792. In *Collected Works.* Vol. 5. Translated by Thomas Saine. New York: Suhrkamp.

Grayling, A. C. 2006. *Among the Dead Cities: The History and Moral Legacy of the WWII Bombing of Civilians in Germany and Japan.* New York: Walker.

Harries, Martin. 2007. *Forgetting Lot's Wife: On Destructive Spectatorship.* New York: Fordham University Press.

Hegel, Georg. 2004. *Introductory Lectures on Aesthetics.* Edited by Michael Inwood and Translated by Bernard Bosanquet. London: Penguin.

Helphand, Kenneth. 2006. *Defiant Gardens: Making Gardens in Wartime.* San Antonio, TX: Trinity University Press.

Hurley, Frank. 1917. *My Diary, Official War Photographer Commonwealth Military Forces.* [...]. Canberra: National Library of Australia, Ms 883.

Kant, Immanuel. 2006. *Anthropology from a Pragmatic Point of View.* Edited and Translated by Robert Louden. Cambridge: Cambridge University Press. First published in German in 1798.

Knell, Hermann. 2003. *To Destroy a City: Strategic Bombing and its Human Consequences in World War II.* Cambridge, MA: Da Capo.

Leopardi, Giacomo. 1982. *Operette Morali: Essays and Dialogues*. Translated by Giovanni Cecchetti. Berkeley, CA: University of California Press. First published in Italian in 1845.

Loader, J. A. 1990. *A Tale of Two Cities: Sodom and Gomorrah in the Old Testament, Early Jewish and Early Christian Traditions*. Kampen: J. H. Kok.

Lowe, Keith. 2007. *Inferno: The Fiery Destruction of Hamburg, 1943*. New York: Scribner.

McConville, J. Gordon. 2012. Hosea, Book of. In *Dictionary of the Old Testament: Prophets*, ed. Mark J. Boda and J. Gordon McConville, 338–350. Downers Grove, IL: InterVarsity Press.

Nossack, Hans. 2004. *The End: Hamburg 1943*. Translated by Joel Agee. Chicago, IL: University of Chicago Press.

Pimlott, John. 1988. The Theory and Practice of Strategic Bombing. In *Warfare in the Twentieth Century: Theory and Practice*, ed. Colib Mcinnes and G. D. Sheffield, 113–139. London: Unwin Hyman.

Sebald, W. G. 2003. Air War and Literature. In *On the Natural History of Destruction*, 1–104. Translated by Anthea Bell. New York: Random House.

Sloterdijk, Peter. 1989. *Thinker on Stage: Nietzsche's Materialism*. Translated by Jamie Owen Daniel. Minneapolis, MN: University of Minneapolis Press. First published in German in 1986.

Sloterdijk, Peter. 2009. *Terror from the Air*. Translated by Amy Patton and Steve Corcoran. Los Angeles, CA: Semiotext(e). First published in German in 2002.

Sloterdijk, Peter. 2012. *The Art of Philosophy: Wisdom as Practice*. Translated by Karen Margolis. New York: Columbia University Press. First published in German in 2010.

Sloterdijk, Peter. 2016. *Spheres, Volume 3: Foams: Plural Spherology*. Translated by Wieland Hoban. Los Angeles, CA: Semiotext(e). First published in German in 2004.

Sloterdijk, Peter, with Hans-Jürgen Heinrichs. 2011. *Neither Sun nor Death*. Translated by Steve Corcoran. Los Angeles, CA: Semiotext(e). First published in German in 2001.

Solnit, Rebecca. 2005. *A Field Guide to Getting Lost*. New York: Viking.

Taylor, Frederick. 2004. *Dresden: Tuesday, February 13, 1945*. New York: HarperCollins.

Thomas, Edward. 1978. Diary of Edward Thomas [1917]. In *The Collected Poems of Edward Thomas*, ed. R. George Thomas, 460–481. Oxford: Clarendon.

Virilio, Paul. 2005. *City of Panic*. Translated by Julie Rose. Oxford: Berg.

Virilio, Paul. 2007. *Art as Far as the Eye Can See*. Translated by Julie Rose. Oxford: Berg.

Weber, Samuel. 2008. *Benjamin's – Abilities*. Cambridge, MA: Harvard University Press.

Wells, H. G., ed. 2005. *The War in the Air*. London: Penguin. First published 1908.

CHAPTER 9

Sublime City and Slimy Slums

Abstract The mixing of the elements of earth, air, fire, and water occurred in the development of modern cities, such as London, another city set in a swamp, portrayed by recent writers and nineteenth-century ones as the sublime city with slimy slums. The lost and repressed slimy swamps return in a fascination with, and horror of, the slimy slums of the city figured and denigrated as jungle or swamp in some late nineteenth-century writers about London. These writers construct a circularity of figurality and so are placist. The pejorative associations of a jungle or swamp are applied to the feral city slum, and the jungle or swamp comes to be seen in the same denigratory (literally 'darkened') light (or dark) as the city slum. Beside(s), or above, the nether world of the uncanny city of dreadful night, nineteenth-century London also possessed and displayed the upper or fore world of the sublime city of delightful light encapsulated in the Crystal Palace built in London for the Great Exhibition of 1851 that fascinated Dostoevsky as the quintessential symbol of modernity. The sublime city is made possible by draining and building on its slimy swamps and it makes slimy slums. All three go hand-in-hand.

The sublime city is founded in slimy swamps, founds slimy slums and becomes monstrous in the Anthropobscene. Such cities include London, Paris, St Petersburg, Berlin, Chicago, Boston, New York, New Orleans, Washington, Toronto, and Venice. The lost and repressed slimy swamps

R. Giblett, *Living with Nature in the Anthropobscene*,
https://doi.org/10.1007/978-3-032-12822-5_9

return in a fascination with, and horror of, the slimy slums of the city figured and denigrated as jungle or swamp in some late nineteenth-century writers about London, such as Arthur Conan Doyle, Friedrich Engels, George Gissing, Jack London, and James Thomson. These writers are what I call placist by constructing a circularity of figurality. The pejorative associations of a jungle or swamp are applied to the feral city slum and the jungle or swamp comes to be seen in the same pejorative light (or dark) as the city slum. The obverse also applies where a human made wet wasteland, such as the city slum or mud hell of World War I, is made to figure a native wetland, such as a swamp or marsh, and the pejorative associations of the former are applied to the latter. This pejorative figuring of wetlands becomes mutually reinforcing and culturally naturalised, but its logic is circular and its ideology is placist.

9.1 London: Marsh Metropolis

The swamp city and marsh metropolis of London dates, like Paris, from pre-Roman and Roman times. Even the name of London might enshrine its wetland beginning as the name may be of Gaelic origin with *lunnd* meaning "marsh" (Ackroyd 2001, 10). London was founded famously, and still sits, on the Thames River. When the city was founded the river was for Peter Ackroyd (2001, 9) "bordered by swamps and marshes." As with most cities set in or beside swamps and marshes, London has a swampy and marshy beginning that the majority of residents do not know about as it has largely been forgotten. The fact that the majority of residents are oblivious to their city's wetlands means that they are present by virtue of their very oblivion, to borrow Benjamin's words in writing about Kafka and the swamp world (Benjamin 1973b, 130; 1999b, 809). London's marshes and swamps have been remembered in Bernardine Evaristo's verse novel *The Emperor's Babe* set in London's Roman and contemporary times with its description of "the marsh saltings/and impenetrable swamps of Thamesmead" (Evaristo 2002, 168–169).

Since Roman times, London grew to become for David Long (2014, 113) "the first city anywhere in the world to reach a population of a million," though for Lehan (1998, 30) London was merely "the first modern city in the Western world" to achieve this distinction. London for Joel Kotkin (2005, 81) was "the world capitalist capital," the centre and seat of power for what Long (2014, 79) claims was "the largest empire the world has ever seen." Eduardo Galeano (2009, 158) described how in the eighteenth century "the world and the sun and the stars revolved around London" with "the splendors of London at the summit of the universe." In the late nineteenth century London for Peter Hall (2014, 32, 51, 426) in his monumental history of urban design and planning was "unchallengeably the greatest city in Europe and even the world," whilst in the early twentieth century it "remained the most interesting, the most vital, the most evidently problematic of all the great cities" and became in the late twentieth century "the world's first true global city." Parisians and Berliners might cavil and contend with some of these partisan views of a long-time resident of London, though they cannot dispute the fact that London owed much of its supremacy to being, for over a century, "the world's busiest port in value and volume of trade" (Hall 2014, 426), a pre-eminence to which the landlocked capital cities of Paris and Berlin could never aspire. London for Manuel de Landa (2000, 80) was "part political capital and part maritime metropolis." It was also part marsh metropolis.

Like Paris and Berlin, London was also a swamp city. The wetlandscape of the swamps and lagoons of the Thames for Ackroyd (2001, 9) has "not entirely faded" as "the mist from the marshes of Westminster destroyed the frescoes of St Stephen's within recent memory." The repressed wetlands of the city return in mist, mud, monsters and metaphors. Monsters are "engendered on slime" in Dryden's words (cited by Ackroyd 2001, 8), just as Satan for Milton was "engendered in the Pythian vale on slime" (*Paradise Lost*, X, 530). Dryden is also referring literally to the crocodiles that once lived there and figuratively to the seditious that now live there. Alligators and crocodiles are the wetland monsters par excellence (as we saw in Chap. 5). Dryden for Ackroyd (2001, 8) "recognised this now forgotten and invisible [wet]landscape of London." As Ackroyd (2011, 1) puts it, "the past is beneath us" down below the surface of the city in its depths. The past is also before us in the names of places. What is beneath and before us can and does resurface and re-present itself.

The repressed wetlands of the city return in the city as monster, "*la monstrueuse cité*" as Verlaine put it in the nineteenth century (cited by Ackroyd 2001, 101). As a monster, the city, as Ackroyd (2001, 2) puts it "consumes and excretes" with "greed and desire." Or should that be gluttony? The city is an orally sadistic monster. The repressed monstrous wetlands of the city return in the monstrous city and in its mud. 'The Modern Age' for Sloterdijk (2012, 165, 167 and 180) is the era of the spatial and technological monstrous with "giant cities, museums and rubbish tips" as "typical contemporary products of the monstrous as an industry." The monstrous modern city of the Anthropobscene is built on monstrous maternal marshes.

The depths of the mud on which the city were built resurface in mud on the surfaces of its streets. Ackroyd (2001, 100, 101) cites various writers from the seventeenth through the eighteenth to the twentieth centuries who complained about London's "soft and stinking mud," a London road that "resembled a stagnant lake of deep mud" and "the thick filth and muck of London."John Gay (2007, 180–181), for instance, in 'Trivia: or, the Art of Walking the Streets of London published in 1716 invokes "Cloacina (Goddess of the Tides/Whose sable Streams beneath the City glide)" in "the nether World" (Gay 2007, 192) (taken up later by Gissing) and leads the reader/walker on a jaunt through the miry mud of London streets (taken up later by Henry Mayhew) to "the slimy shore" of the Thames (Gay 2007, 191). For Clare Brant (2007, 110; see also Jenner 2007) "mire is dirt" and for Mary Douglas (cited by Brant 2007, 110) "dirt is matter out of place." Gay for Brant (2007, 109) "sees the waterways and streets as unclearly divided [...] Mud collapses a binary of earth and water." In Douglas's terms, in mud and with it, earth is in the place of water and water in the place of earth. Another name for this category and substance interpellated between the solid and the liquid is slime (as we have seen previously). The city is not like the God of Genesis 1:2 who clearly divided land from water. Mud is wet land that collapses the binary presence of earth (+ 1) into the lack and absence of water (0) and thereby collapses (literally) the relationship between them so that water becomes the presence/present and earth is absent and lacking. Mud and flood rhyme, as do sublime and slime.

A poem published in *Punch* in 1855 personified the river Thames lying in a bed of slime:

> King Thames was a rare old fellow,
> He lay in his bed of slime,
> And his face was disgustingly yellow,
> Except where 'twas black with slime.
> Hurrah! Hurrah! for the slush and slime! (cited by Melosi 2008, 37)

The wetland past of the city of London returns also in science fiction about the future, such as in Richard Jefferies's *After London* and J. G. Ballard's *Drowned World*. In both novels, London reverts to maternal marsh (as we will see in the following chapter). P. D. Smith (2012, 327) argues that in *After London* "the largest city in the world was not too big to be swallowed up by nature." Smith figures nature as an orally sadistic monster. More precisely, in *After London* the city of London is swamped by nature. As for Gay in the eighteenth century, and as Margaret R. Hunt (2007, 120) puts it, "London itself is figured as both feminine and dangerous," a *femme fatale*, rather than as the Great Mother Goddess who is feminine, dangerous and nurturing - just as wetlands are.

Gay's poem *Trivia*, as Hunt (2007, 120) points out, "takes its name from a minor Roman goddess whom John Gay fancifully appoints to be patroness of streets and highways." Gay trivialises carnivalesquely the city, including the goddess Cloacina, to celebrate the grotesque lower bodily and urban strata. He does not slimily celebrate the city as founded in and on the monstrous Great Mother Goddess of swamps (unlike Jefferies who does so later). Ballard's and Jefferies's visions of the past and future have present pertinence for without the Thames Barrier, as Smith (2012, 328) points out, "the Thames flood plain would revert to marshland" and "Nature would have finally reclaimed the [wet]land on which the once great city had been built." The wetland reclaimed by the city would have been re-reclaimed by the wetland. Return of the repressed indeed.

Auguste Barbier's 1837 poem 'London' figures the city both as a cathedral of industrial capitalism and as a monster:

> Long black chimneys, the steeples of industry,
> Open their mouths and exhale fumes
> From their hot bellies to the open air;
> Vast white domes and Gothic spires
> Float in the vapour above the heaps of bricks. (cited by Benjamin 1999a, 452)

"An ever swelling, unapproachable river" is also figured as a monster "rolling its muddy currents in sinuous onrush/Like the frightful stream of the underworld," the Styx. "Above the sky tormented, cloud upon cloud,/ The sun, like a corpse, wears a shroud on its face." Just as "A great tide polluted and always unsettled/Recirculates the riches of the world," the sun is obscured by air pollution and becomes the figure of death. The black sun of mourning and melancholy is produced here, not the black airs of the black waters of the wetland, but by the polluted air of the city that took the wetland's place in the Anthropobscene.

The repressed wetlands of the city return in tropes. For many late nineteenth-century writers, such as Friedrich Engels, George Gissing, Jack London, and James Thomson, the modern, industrial city (principally London for these writers) has a dark underside figured as stagnant pool, bottomless abyss, nether world, or dreadful night. London for Smith (2012, 265 and 297) is "the first metropolis of the industrial age" and "the capital of the industrial world," certainly in the nineteenth century. The commonplace responses of dread and horror that were projected on to the native quaking zone of the swamp were displaced on to the industrial quaking zone of the urban underside. These writers are 'placist' in that they ascribe characteristics to a (man-made) place (the city) that were previously ascribed to a place not made with human hands (jungle, abyss, nether-land, swamps, etc.). The negative connotations that attach initially to the native quaking zone are attached subsequently to the feral quaking zone.

It is a commonplace of detective fiction that the modern city has a dark underside and that the role of the detective is to penetrate the darkness and bring it to light, and to the light. The first writers of the genre also construed the modern city as swamp and sewer and the detective as drainer and sanitary engineer. Conan Doyle (1981, 15) in *A Study in Scarlet*, his first Sherlock Holmes story, figures London as "that great cesspool" into which the wastes and wasters of the British empire ran. The role of the detective is to wade through this pool of shit and urine, and drain it. Doyle also figured the city as wilderness and the detective as pioneer who crosses the frontier between civilisation and wilderness. In the same novel Conan Doyle (1981, 16) refers to "the great wilderness of London." The role of the private detective is to first cross the frontier between civilisation and barbarism, enter the wilderness and then civilise and tame the wilderness. In one of his later Sherlock Holmes stories Conan Doyle (1981, 488) refers to "the dark jungle of criminal London." The role of the private detective is to slash through the tropical growth and penetrate the darkness of the modern city, and bring enlightenment to its benighted citizens in the Anthropobscene.

In another of his Sherlock Holmes stories Conan Doyle (1981, 636) refers to "the opalescent London reek." Given the state of the air, it is hardly surprising then that Conan Doyle (1981, 98) in *The Sign of Four* has Watson describe an autumn evening in London in typical, if not clichéd, nineteenth-century terms:

> a dense drizzly fog lay low upon the great city. Mud-coloured clouds drooped sadly over the muddy streets. Down the Strand the lamps were but misty splotches of diffused light which threw a feeble circular glimmer upon the slimy pavement. The yellow glare from the shop-windows streamed out into the steamy, vaporous air and threw a murky, shifting radiance across the crowded thoroughfare. There was, to my mind, something eerie and ghost-like in the endless procession of faces which flitted across these narrow bars of light—sad faces and glad, haggard and merry. Like all humankind, they flitted from the gloom into the light and so back into the gloom once more. I am not subject to impressions, but the dull, heavy evening, with the strange business upon which we were engaged, combined to make me nervous and depressed.

This view of the modern city from St Petersburg to London can be found from Bely to Eliot, especially with fog. As with many writers, Doyle emphasises the uncanniness of the modern city. Like Blake who in his archetypal poem 'London' "mark[s] in every face I meet / Marks of weakness, marks of woe," Doyle reads the physiognomy of faces in the street.

T. S. Eliot may have had Dostoevsky and his Petersburg in mind when he called London "unreal city"; he may also have had Doyle and his London in mind when he personifies, or more precisely felinofies, the fog as a cat in 'The Love Song of J. Alfred Prufrock':

> The yellow fog that rubs its back upon the window-panes,
> The yellow smoke that rubs its muzzle on the window-panes
> Licked its tongue into the corners of the evening,
> Lingered upon the pools that stand in drains

Fog has a life of its own, a feline life that is made animal in Eliot's vision of the unreal city and thereby ceases to be human, though the fog of the microclimate of cities is human made.

In the nineteenth century Nathaniel Hawthorne also referred to "the black heart of London" (cited by Ackroyd 2001, 102). Darkness for Ackroyd (2001, 103) is of "the city's essence [...] London is possessed by darkness." London, to borrow from Joseph Conrad, is "the heart of darkness." The heart of darkness is not only in the rogue colonist Kurtz, nor only in the deepest, darkest jungles of the "dark continent" of Africa, nor of female sexuality for European men as Freud thought, but in the heart of the colonising capital as Marlow in Conrad's *Heart of Darkness* said of the Thames "this also has been one of the dark places of the earth" (cited by Ackroyd 2001, 542). Ackroyd goes on to suggest that "the derivation of its very name, pre-Celtic in origin, is *tamasa*, 'dark river'." Ackroyd uses Conrad's title of his novel for chapter 6 of *London Under* about London's sewers, the Thames beneath the city, the subterranean rivers of shit (Ackroyd 2011, 65–87). Some of these sewers have been rivers. For Long (2014, 9) London has many "lost rivers," "several of which are now little more than sewers." Even the Thames above winding its way through the city was not much better. The Thames for Long (2014, 60) is, or has been, "the city's great open sewer." This is perhaps not quite what the original namers had in mind when they called it the dark river.

For other late nineteenth-century writers the dark underside of the city looms large in slightly different, though similar, ways to Doyle, Hawthorne and Conrad. For Jack London it is the abyss; for Rudyard Kipling and James Thomson it is 'The City of Dreadful Night'; for George Gissing it is the nether world; and for Friedrich Engels it is a stagnant pool. This figuration and association persists into some of the most recent writing about the city, such as Antoine Picon's reading of the anxious landscape of the rusting city. How this dark underside is figured adumbrates the politics, pleasures and fears of city places and urban spaces (Picon 2000).

The darkness of the modern city is constituted not only by its immoral underside, its dens of thieves, its parade of prostitutes, its labyrinthine slums, but also by the physical and moralised underside of the open drains or closed sewers that service it and the swamps on which it may have, more than likely, been built. The latter is often employed to figure the former with the *demi-monde* of the slums figured as swamp, or sewer, or muddy shore. Henry Mayhew's aim for what he calls "the neglected class" is to try to "lift them out of the moral mire in which they are wallowing" (Mayhew 1985, 41). Mayhew's source for his miry metaphor is not only John Gay's *Trivia* but also his own description later of the "mud-larks" who "plash their way through the mire" (Gay 2007, 210) and wade "through the mud left on the shore by the retiring tide" and whose "bodies are grimed with the foul soil of the river" (Gay 2007, 209). Both sides of the figuration come together in the uncanny and the nether world (or nether lands).

9.2 Nether World

In George Gissing's 1889 novel, *The Nether World* the slums of London nurture "a rank, evilly-fostered growth," like a poisonous swamp plant in contemporaneous representations of swamps and its vegetation. His narrator is not referring to a plant here though. He is comparing one of the working-class girl characters, "not to some piece of exuberant normal vegetation," but to a kind of festering vegetative sore on the backside of the body politic. Yet their labour is necessary for the functioning of the body politic and could not survive without it.

The narrator offers the opinion that "the putrid soil of that nether world yields other forms besides the obviously blighted and sapless" (Gissing 1992, 8). Indeed, it yields the bad smelling and the morally bad. This is hardly surprising when "filth, rottenness, evil odours, possessed these superfluous dens of mankind and made them gruesome to the peering imagination" (Gissing 1992, 74). Into "the jaws of this black horror" the narrator, or at least his imagination, has peered. The nether world is figured as monster and the narrator as dentist who peers (albeit in imagination, not in reality) into its orally sadistic jaws. The nether world is a living creature to be shunned as an abomination. The narrator leaves us in no doubt about this when he describes "a disagreeable quarter, a street of squalid houses, swarming with yet more squalid children" (Gissing 1992, 129). These swarms of children have developed from a "swarm of babies" (Gissing 1992, 132). According to the Levitical interdiction, swarming creatures are an abomination and are to be abhorred (as we saw in Chap. 2).

The swarming nether world of the city with its vertiginous depths is worlds away from the pastoral upper world of the country with its flat horizons (though it too has its dark under side). This contrast comes into stark relief when making a journey by railway from the city to the country:

> Over the pest-stricken regions of East London, sweltering in sunshine which served only to reveal the intimacies of abomination; across miles of a city of the damned, such as thought never conceived before this age of ours; above streets swarming with a nameless populace, cruelly exposed by the unwonted light of heaven; stopping at stations which it crushes the heart to think should be the destination of any mortal; the train made its way at length beyond the outmost limits of dread, and entered upon a land of level meadows, of hedges and trees, of crops and cattle. (Gissing 1992, 164)

The unhomely and unhealthy artificial wet land of the city for Gissing gives way to the homely and healthy natural dry land of the country:

> It is merely one of those quiet corners of flat, homely England, where man and beast seem on good terms with each other, where all green things grow in abundance, where from of old tilth and pasture-land are humbly observant of seasons and nations, where the brown roads are familiar only with the tread of the labourer, with the light wheel of the farmer's gig, of the rumbling of the solid wain. (Gissing 1992, 164–165)

This is an idealised picture straight out of Constable of an organic community with its pre-industrial technologies living in pastoral harmony in accord with seasonal rhythms and where the growths are green and abundant, not rank and evilly festered, nor blighted and sapless. It nevertheless has its dark underside as Gissing ([1889] 1992, 165) goes on to relate, how, "here, as elsewhere, the evil of the times was pressing upon men and disheartening them from labour. Farms lying barren, ill-will between proprietor and tenant, between tenant and hind, departure of the tillers of the soil to rot in towns that have no need of them."

The normal, fertile country has been blighted and the symptoms of rural depopulation are blamed on the age, rather than its causes in industrial capitalism diagnosed. The country still has its consolations, though, as a salve for the deprivations of the city: "Danbury Hill, rising thick-wooded to the village church, which is visible for miles around, with stretches of heath about its lower slopes, with its far prospects over the sunny country, was the pleasant end of a pleasant drive" (Gissing 1992, 165).

The church is a symbolic beacon on the hill in dark times, a symbol of the city of God on high, and a light at the end of the tunnel to which the traveller journeys and from which the viewer can command a pleasing prospect of the country. This pleasing prospect is unlike the displeasing prospects of the city where the traveller walks "through all the barren ways and phantom-haunted refuges of the nether world" (Gissing 1992, 247). Both country and city have become barren, but at least the country is not the city of the dammed and the dead. "Mad Jack" declares that "'This place to which you are confined is Hell… This is Hell – Hell – Hell!'" (Gissing 1992, 345). Similarly in Arthur Morrison's (1996, 2) *A Child of the Jago* the rejoinder to the stock epithet "go to hell" uttered by one character is "'Hell? You're in it … there can be no hell after this'" uttered by another character. Similarly for the semi-eponymous Shelley, "hell is a city much like London" (cited by Benjamin 1973a, 59).

The pleasing prospect from Danbery Hill is worlds away from Shooter's Gardens whose "walls stood in perpetual black sweat; a mouldy reek came from the open doorways; the beings that passed in and out seemed soaked with grimy moisture, puffed into distortions, hung about with rotting garments" (Gissing 1992, 248). The verdant horizontal surface of "England's green and pleasant land" is contrasted with the vile, vertical walls of the Farrington Road buildings:

> Vast, sheer walls, unbroken by even an attempt at ornament; row above row of windows in the mud-coloured surface, upwards, upwards, lifeless eyes, murky openings that tell of bareness, disorder, comfortlessness within [...] An inner courtyard, asphalted, swept clean – looking up at the sky as from a prison. Acres of these edifices, the tinge of grime declaring the relative dates of their erection; millions of tons of brute brick and mortar, crushing the spirit as you gaze. (Gissing 1992, 274)

This is not Benjamin's Berlin courtyard of his childhood "where the city opened itself to the child," but the prison house of modernity closed in upon itself, though both are full of intimate, physical possibilities (Benjamin 1979, 330–331; 1999b, 623). These possibilities are exciting sexual ones for Benjamin, but distressing disease-laden ones for Gissing for "the air was poisoned with the odour of an unclean crowd" (Gissing 1992, 274). Gissing subscribed to the miasmatic theory of disease commonplace at the time in which bad air (literally malaria) caused this and other diseases.

9.3 City Conditions

Engels subscribed to the miasmatic theory too. In his 1892 'Preface to the English Edition' of his *Condition of the Working-Class in England* Engels (1987, 42) quotes an article of his own in which he had said that the East End of London in 1845 was "an ever-spreading pool of stagnant misery and desolation." In short, it was an industrial quaking zone of misery and desolation in the Anthropobscene. He is pleased to report 40 years later that "that immense haunt of misery is no longer the stagnant pool it was six years ago" (Engels 1987, 45). Misery has not ceased, but at least the East End is no longer the stagnant pool it was - due in his mind to unionism. Yet the "great towns" of industrial capitalism are stagnant pools not merely metaphorically and morally, but also literally and materially as "the streets are generally unpaved, rough, dirty, filled with vegetable and animal refuse, without sewers or gutters, but supplied with stagnant pools instead" (Engels 1987, 71). These stagnant pools in cities are worlds away from Thoreau's (1962, *VII*, 304) "stagnant ponds or pools in woods floored with leaves." From Engels's (1987, 71) "filthy streets," as would be expected, "a horrible smell" arises. The Aire River of Leeds, "thick, black, and foul, smelling of all possible refuse," "engenders miasmatic vapours" (Engels 1987, 81) according to a journal reporting the sanitary conditions of working people in cities from which Engels quotes. Similarly in Manchester:

> at the bottom flows, or rather stagnates, the Irk, a narrow, coal-black, foul-smelling stream, full of debris and refuse, which it deposits on the lower right bank. In dry weather, a long string of the most disgusting blackish-green slime pools are left standing on the bank, from the depths of which bubbles of miasmatic gas constantly arise and give forth a stench unendurable. (Engels 1987, 89)

Hardly surprisingly Engels concludes that Manchester is "this hell upon earth" which "arouses horror" (Engels 1987, 92).

Miasmatic gas is often described in short, and in a word, as effluiva that arise from "masses of refuse, offal and sickening filth [that] lie among standing pools in all directions" with the result that "the atmosphere is poisoned by the effluvia from these" (Engels 1987, 98). Although Engels does not belabour the miasmatic theory of disease, he does not demur from repeatedly quoting the journal reporting the sanitary conditions of working people in cities to the effect that in such streets "a mass of dried filth and foul vapours are created, which not only offend the [senses of] sight and smell, but endanger the health of the inhabitants in the highest degree" (Engels 1987, 78). Later he more explicitly argues that "the filth and stagnant pools of the working people's quarters in the great cities have, therefore, the worst effect upon the public health, because they produce precisely those gases which engender disease; so, too, the exhalations from contaminated streams" (Engels 1987, 128–129). Certainly, as Engels says, "life in large cities is, in itself, injurious to health," but whether this can be attributed to what he calls "gases decidedly injurious to health" (Engels 1987, 128) is another question and more a matter of nineteenth century misdiagnosis. Nevertheless, the city is a place of death for many nineteenth-century writers. For Thomson (1993, 29) "The City is of Night; perchance of death." Following in a similar vein Kipling is in no doubt in his short story 'The city of dreadful night' published in 1891 that "the city was of Death as well as Night."

The stagnant pool in the city made with human hands is a deadly place. The city often had its beginnings in the stagnant pools of swamps not made with human hands. Many modern cities, or areas in them like the East End of London, for Mike Davis (2006, 82) "the Victorian world's greatest slum," were built on reclaimed marshes, or drained swamps. As Robert Neuwirth (2006, 179) says, "all cities start in mud." Cities begin in mud and end in crud. Mud for Thoreau (1962, *V*, 499) writing in his journal in 1853 is "Nature's womb." All cities start in nature's womb of mud and end up in culture's tomb of crud. Crud is culture's tomb. The city's end is in crud. The city is born from mud and dies in crud; it is born from nature's womb and dies in culture's tomb. The city made of mud-bricks or concrete, a mixture of the three elements of water, earth and air, or of clay-bricks baked in fire, a mixture of four elements, ends up with the elements falling apart and returning to dust. Yet cultural crud is often figured as natural mud. The filled or drained swamp on which the city was built, its spatial, geographic and historical repressed, returns in the fascination with the dark underside of the stagnant pools of the city as metaphors for the city. James Thomson (1993, 29) describes in *The City of Dreadful Night*, his meditation on the modern city (obviously based on London), how:

> A river girds the city west and south,
> The main north channel of a broad lagoon,
> Regurging with the salt tides from the mouth;
> Waste marshes shine and glisten in the moon
> For leagues, then moorland black, then stony ridges;
> Great piers and causeways, many noble bridges,
> Connect the town and islet suburbs strewn.

The river is a monster that girdles the city and regurgitates the black waters of waste wetlands.

9.4 London on London

The city makes a mark on the earth—in space— and marks a place in history—in time. Its dwellers, its citizens, make a mark on its streets. When Jack London (2001, 39) was in London in 1902 he remarked repeatedly on "the slimy, spittle-drenched sidewalk." He contrasts "the solid walls of bricks" with "the slimy pavements, and the screaming streets" (2001, 4), the solid and vertical contrasted with the becoming liquid and horizontal. In this spatial poetics, if not metaphysics, the solid and vertical is valorised over the slimy and horizontal.

London had just witnessed "a market [where] tottery old men and women were searching in the garbage thrown in the mud for rotten potatoes, beans, and vegetables, while little children clustered like flies around a festering mass of fruit, thrusting their shoulders into the liquid corruption, and drawing forth morsels but partially decayed which they devoured on the spot" (2001, 3–4). This urban, industrial swamp of decomposing and stinking vegetable matter is for London "that wilderness" (2001, 4). Similarly for Morrison (1996, 10) "the whole East End was a wilderness of slums." Perhaps it was not unlike the swamp of the wilderness itself with their similar smells to the undiscerning nose to which all bad smells are bad in the same way. "Filth and noisomeness" are coupled together for Mayhew ([1861–1862] 1985, 109 and 177) too; bad matter and bad smells go together. Dirt as matter out of place is also noisome as the bad smell of a place. Whether that place was a slum or a swamp did not make much difference.

The slums of London (in 2 senses) are a site of horror as Jack goes on to state that "for the first time in my life the fear of the crowd smote me. It was like the fear of the sea, and the miserable multitudes, street upon street, seemed so many waves of a vast and malodorous sea, lapping about me and threatening to well up and over me" (London 2001, 4). This fear of engulfment by the liquid and horizontal is evinced not only by the sight, sound and touch of the crowd and the city, but also by its smells. Yet London finds that "when at last I made into the East End, I was gratified to find that the fear of the crowd no longer haunted me. I had become a part of it. The vast and malodorous sea had welled up and over me, or I had slipped into it, and there was nothing fearsome about it" (7).

The crowd takes on some of the qualities of the urban swamp as a few pages later London describes "a woman of the finest grade of the English working-class, with numerous evidences of refinement, being slowly engulfed by that noisome and rotten tide of humanity which the powers that be are pouring eastward of London town" (13–14). The "strange, vagrant odours" (122) are like the vagrants who frequent the vagrant space of the slum. The human tide is flotsam and jetsam. The festering mass of fruit is ultimately indistinguishable from "the festering contents of slum, stews and ghetto." This human sewage "resemble[s] some vile spawn from underground" (87). This human sewage is not spawned so much from under*ground* but from what London (2001, vii) calls "the under-world of London," the city, and the under-world of London the writer, whose tropes give birth to the denizens of the slum. London (2001, xiii) "went down into the underworld" of his eponymous city "with an attitude of mind which I may best liken to that of the explorer." He penetrates into the urban underworld of the slum swamp just as the explorer in the colonies descended into the underworld of the swamp. Just as the romance hero emerged with his manhood vindicated through trial and tribulation, so does London (the writer) emerge to write his book about London (the city and himself) and preface it in Piedmont, California, literally at the foot of the mountain, a dry, elevated place thousand of miles away from, and thousand of feet above, the wet, low place of the urban underworld of the East End of London (the city). Like Dante, he descends into "the inferno" (London 2001, 1, 6) and "infernal regions" (London 2001, 52, 68) of "hell on earth" (London 2001, 68) and the inferno of London (London 2001, 159) to ascend into the paradise of heaven on earth of God's mountains in California as John Muir called them.

Similarly Arthur Morrison (1996, 1) begins *A Child of the Jago* with a description of:

> the narrow street all the blacker for the lurid sky; for there was a fire in a farther part of Shoreditch, and the welkin was an infernal coppery glow. Below, the hot, heavy air lay a rank oppression on the contorted forms of those who made for sleep on the pavement; and in it, and through it all rose from the foul earth and the grimed walls a close, mingled stink—the odour of the Jago.

For Morrison, the Jago, "for one hundred years the blackest pit in London lay and festered," like an open, infected wound inflicted on the body of the earth. In the centre of the blackest pit was Jago Court, "the blackest hole in all that pit." This is the centre of the inferno where Satan resides in Dante's Inferno with the denizens of the slum in this secular theology, a world where God is dead. The sublime is God in secular theology and slime is Satan.

Like Morrison, London, the urban and colonial explorer, the endo- and exo-colonist, the colony within at home and the colony without away from home, is apt to find both decay and decomposition wherever he looks look and tread. Although London and Morrison did not find miasma and malaria (unlike his colonial counterparts or previous urban explorers), Morrison (1996, 2) noted "the stifling air" and London (2001, 23) discovered "disease germs that fill the air of the East End" and "the effluvia and vile exhalations of overcrowded and rotten life" (London 2001, 126). He also found that "rottenness is a slimy desecration of the sweetness and purity of nature" (London 2001, 87). The denizens of "city slime" (London 2001, 113) are slimy creatures of slimy pavements, "a fearful slime that quickened the pavement with life" (London 2001, 151). Human life has taken on a vegetable, animal and liquid life of its own. Humans are not autochthonous creatures that spring from the earth, or swamp, but creatures that spring from the urban swamp, the slum—they are auto-ex-metropolis, self-generating from the city. For London (2001, 28), "a spawn of children cluttered the slimy pavement, for all the world like tadpoles just turned frogs on the bottom of a dry pond," or more precisely on the bottom of a dried-up industrial urban wet wasteland in the Anthropobscene. The categories of solid and liquid, animal and vegetable, and the elements of earth, water and air are all mixed up. "The place swarmed with vermin" (London 2001, 83), with "swarming children" (London 2001, 91) and with workers who swarm (London, 2001, 113). In Morrison's (1996, 1) *A Child of the Jago*, in Jago Court "the human population swarmed in thousands." Swarming creatures neither walk, nor swim, nor fly, but are hybrid creatures, or monsters, who do all three.

As swarming creatures, they are neither fish, flesh, nor fowl and they fall under the Levitical interdiction of being an abomination. For London (2001, 152), "they are a new species, a breed of city savages [...] The slum is their jungle." This jungle is not a tropical rainforest where nature is green in leaf and branch, but an urban jungle where culture is red in tooth and claw. For London it would be "far better to be a people of the wilderness and desert [...] than to be a people of the machine and the Abyss" (London 2001, 153). Better to be creatures of the swamp and jungle than of that urban swamp and jungle called the slum; better to be creatures of the living waters of wetlands than to be minced to death in "the Abyss [which] is literally a huge man-killing machine" (London 2001, 23) just as the slimy trenches of World War I were later; better to be a living swamp creature than to be "the living deaths" (London 2001, 152) at "the bottom of the Abyss" (London 2001, 152); and better to be in the jungles of "Darkest Africa" (London 2001, 1) than in "the jungle of empire" (London 2001, 149).

The landscape of urban slums is not only prescient of the landscape of trench warfare, but also the product of class warfare perpetrated by the military-industrial ruling capitalist upper class against the lower classes. For London "here, in the heart of peace, is where the blood is being shed" (London 2001, 135). "In London the slaughter of the innocents goes on on a scale more stupendous than any before in the history of the world" (London 2001, 150), including Herod's in the wake of the birth of Jesus (London 2001, 135). This slaughter is perpetrated by "the men of England, masters of destruction, engineers of death [...] men of steel [...] war lords and world harnessers" who have "mastered matter and solved the secrets of the stars" (London 2001, 76). London also maintained that if the masses of the slums attempted revolt they would perish "before the rapid-fire guns and the modern machinery of warfare" (London 2001, 123). Rather than revolting, they ended up as cannon fodder in the 'meat-mincer' of the muddy fields of Flanders during World War I.

The denizens of the East End are monstrous like the city they inhabit (hardly live in). The city takes on a life of its own as it is what London calls "the monster city" (London 2001, 14) with its own aberrant and peculiar anatomy. Night in the city is not merely the time of sunlessness but "the black night of London [which] settles down in a greasy pall" (London 2001, 14). London for James Thomson is 'the City of Dreadful Night,' but not for Jack London. The East End is "often called," according to London (2001, 113), "the City of Dreadful Monotony." Just as wilderness swamps were often called dreadful and monotonous, so the urban swamp is invoked in similar terms, though for London (2001, 114) "the East End does merit a worse title. It should be called 'The City of Degradation.'" It takes one lower into the depths and bowels, not of the good earth and wilderness swamp, but of the evil city and city swamp, "the perilous depths of the East End" as Morrison (1996, 10) puts it. In this body politics of the city the park is not for London (2001, 31) "a lung of London," which it should be in accordance with the dictates of the desirable Victorian body politic, but "an abscess, a great putrescent sore" on the bottom or backside of the body politic that should be treated and the patient cared for, not lanced in the gentrification of slum clearance as Hausmann did in Paris.

The ways in which various writers describe and figure the city play out theirs fears.

James Thomson figured the dark underside of the late nineteenth century city as dreadful night and Engels as a haunt of misery. Antoine Picon (2000) in the current century characterised this side of the late twentieth/early twenty-first century city as an "anxious landscape." Picon contrasts Manhattan as "a magic city made of crystal" and as "the celestial Jerusalem" with "the hell, or purgatory" of Newark with its "creeping swamps" where everything is "rusted out" and "irreparably polluted yet somehow endowed with a strange beauty." In Yeatsian terms, a terrible beauty is born in the rusted city. The central business district with its sublime skyscrapers is heaven compared to the slimy hell of industrial swamps in the Anthropobscene. The slimy city is the counterpoint and counterpart to what Norman O. Brown (1959, 281–283) calls "the city sublime."

Picon is working in the tradition of Blanchard Jerrold and Gustave Doré (2005, 11) and their depiction of nineteenth-century London, in particular of "the dead shore" of the Thames River with "the muddy, melancholy bank" with "rust upon everything." As the marsh has traditionally between associated with melancholy, Jerrold's and Doré's pilgrimage could have referred to the muddy, melancholy bank as "a slough of despond" to make an explicit connection with John Bunyan's *Pilgrim's Progress*. Picon does not use such archaic language, but the implication or impression is that the Newark industrial swamps are depressing, though strangely beautiful, places.

Similarly when Jerrold and Doré (2005, xxviii) went on their "travel[s] about London in search of the picturesque" they "discovered that it abounded [...] in picturesque scenes." The picturesque is not an object, nor a feature of the object, but a point of view, or a modality that sets up a relation between the subject and object, the viewer and the scene and seen. If one goes in search of the picturesque, one will invariably find it because one carries the picturesque point of view with one. For Jerrold (2005, 4, 5), "the lesson which Doré's pictorial renderings of our mercantile centre will teach [...] is that London, artistically regarded, is not, as the shallow have said so often, an ugly place, given up body and soul to money-grubbing." Yet there are limits to the picturesque as Jerrold (2005, 15) concedes that in the poverty of the poor "there is nothing picturesque." Poverty cannot be depicted in picturesque terms because the picturesque is an impersonal point of view from a distance whereas poverty is personal (see Doré's illustration no. 119, 'Wentworh St, Whitechapel'; Jerrold and Doré 2005, 145). The poverty of the poor is conveyed much better in Jerrold's words used by John Coolidge (1994, 53) as a caption to this illustration, than in Doré's drawing: "beat-up alleys with pools of water. Here, tattered young women [...] look at you sullenly, fiercely, what suffering, what hunger can be read in their meager bodies, hungry since the day they were weaned." Jerrold does not reflect on the inability of the picturesque to convey the poverty of the poor and the ability of language to do so. The picturesque is also unable to convey the poverty of impoverished landscapes. Rusting industrial ruins and other wastelands have become an object of contemporary fascination, and fashion in the Anthropobscene (again, see the cover photo of *Living with Nature in the Anthropobscene*, both symptom and talisman).

9.5 Slums

The dark side of both the late nineteenth century and late twentieth/early twenty-first century cities are invoked in terms of the swampy and uncanny and testify to a fascination with, and horror of, the urban nether world and the (in this case, artificial) quaking zone known otherwise and simply as "the slum" or "squatter communities" as Neuwirth (2006, 241) prefers to avoid the pejorative connotations of slum, or "squatter settlements" as Kotkin (2005, 132) prefers. The slum in the early twentieth-first century is the habitat for one billion people. For Davis (2006, 19) "instead of cities of light soaring toward heaven, much of the twenty-first-century urban world squats in squalor, surrounded by pollution, excrement and decay." Squats, in other words, in a man-made swamp, and not only squats physically on its haunches, but also illegally on land its residents do not own (Neuwirth, 2006, 12). Squatting today for Davis (2006, 39) "continues primarily in low-value urban land, usually in hazardous or extremely marginal locations, such as floodplains, hillsides, swamps or contaminated brownfields." The slum for Davis (2006, 22) was "first and above all envisioned as a place where an incorrigible and feral social 'residuum' rots in immoral and often riotous splendour." Vegetable decomposition figures moral decay and anarchic display. The slum is the arsehole or cloaca of the body politic, a festering wound inflicted on the body of the earth. Whether or not it can, or how it should, be treated is another matter.

For Davis (2006, 95) "urban inequality in the Third World is visible even from space: satellite reconnaissance of Nairobi in Kenya reveals that more than half of the population lives on just eighteen percent of the city area." This area includes Kibera, "Africa's largest mud hut metropolis" (Neuwirth 2006, 70), or slum. Satellite reconnaissance of Nairobi also reveals nearby Nairobi National Park, the world's largest National Park within the municipal bounds of a city. Two iconic and interdependent features of modernity sit cheek-by-jowl with each other: nationalism and landscape preservation and display in the national park side-by-side with the underbelly of urbanity in the slum; the national park with vast open areas populated sparsely by rangers and tourists with the slum crowded and over-built with rusting iron shanties and muddy, sewery paths; and the national park created by the dispossession and removal of its native owners and inhabitants into the slum.

The British for Davis (2006, 52) have the dubious distinction of being "the greatest slum builders of all time." They designed and built railways (or at least their labourers did) and the railways, in turn, for Coolidge (1994, 32) were "the greatest creators of London slums." British railways, at home and abroad, created the interstices between railway lines that were filled by slums. The communication and transportation technology of the railway excommunicated slum-dwellers between and beside its lines. Nairobi, for example, come into being in 1899 when the British wanted to span East Africa with a railway and used a small Masai settlement at the confluence of several small rivers as the staging point for the construction of the railway (Neuwirth 2006, 91).

Railways created the slum during the period from the nineteenth century to the early twentieth century. Road transport and satellite communication are the greatest creators of planetary slums today that excommunicate slum-dwellers between its super-highways and vectors. Yet the sanitary conditions are pretty much the same. Many slum-dwellers today live in what Davis (2006, 144) calls "the sanitary equivalent of the mud hell of World War I trench warfare." The slum is the product of class war fought against the lower classes. Intra-urban class warfare produced a landscape very similar to that produced by international armed warfare. Thirty years before Davis, Fussell 2000, 149) had noted the converse: "the similarity of the trench scene to modern, urban, industrial squalor." It is hardly surprising that both places end up looking and being the same as both were the product of class warfare fought against the earth, against both the lower earthly and class strata.

One writer to portray graphically and poetically the horror of the "mud hell" of World War I trench warfare is Edward Thomas. In his diary he writes that this landscape is characterised by "filth" (Thomas 1978, 461) and how it is a place of "death living" (not a place of living death, nor of the living dead (Zombies), but dead living) (461). It is a place of "dirt and depredation" (471) as it is "muddy" (471), which is either "stiff deep mud" (477) or "muddy and slippery" (481). It is an artificial, 'man-made' wetland made out of marshes. It is also a place of ruins as buildings are ruined and occasionally "sordid ruin" (472).

The tone of the sky is perpetually, remarkably and repeatedly "dull" (on just about every day). The sky above and the wet land below occasionally combine into "a dull muddy day" (471) and a "dull marsh" (476). Yet the sky is also a site of strange beauty as an "enemy plane [is] like [a] pale moth beautiful among shrapnel bursts" (469) and there are "lovely white puffs of shrapnel round planes high up" (470). The overall impression and depiction, however, is of a waste land, a "snowy broken land with posts and wires and dead trees" (469) and "ghastly trees and ruins" (474). All in all, it is a ruined swamp with "skeletons of whole trees" (477) with "waste trenched ground" (477) and "shell holes full of blood stained water" (477). It is hell: "some day this will be one of the hottest places this side of Hell, if it is this side" (472).

The city founded in war, or at least of the preparations for it as Virilio and Lotringer (1983, 3) put it, becomes the landscape of warfare, whether it is the landscape of World War I trench warfare or the landscape of World War II aerial warfare. "The urban institution of war," as Lewis Mumford (1961, 42) puts it, makes the city into the landscape of warfare. This landscape is either the landscape of class and trench warfare as in WWI or the landscape of international and aerial warfare as in WWII. What Mumford (1961, 53) following Geddes calls "a living urban core, the polis, ends in a common graveyard of dust and bones, a Necropolis, or city of the dead: fire-scorched ruins, shattered buildings, [...] heaps of meaningless refuse"—in a word crud, as in the aftermath of the fire-bombing and firestorms of Dresden, Hamburg, Wurzburg, and Tokyo using what Smith (2012, 182) "the new technology of urban mass destruction [of] aerial bombing." The city is our greatest creation, but also our greatest site of destruction by one of our greatest destructive technologies in bombers. The city begins in the natural mud of the wetland, goes through the cultural mud of the slum and ends in the cultural crud of the wasteland in the Anthropobscene.

9.6 Crystal Palace

Beside, or above, the monstrous city of modern industry and the nether world of the uncanny city of dreadful night, nineteenth-century London also had the upper or fore world of the sublime city of delightful light encapsulated in the Crystal Palace built in London for the Great Exhibition of 1851 (Sloterdijk 2016, 321). As the arcades were for Benjamin the representative architectural monument of modernity, so is the Crystal Palace for Dostoevsky, Berman, and Sloterdijk. The crystal palace for Dostoevsky's underground man represents a place where suffering is unthinkable and doubt is impossible as "suffering is doubt, negation–what kind of crystal palace would it be if one could doubt in it?" (Dostoevsky 2009, 32) The crystal palace is a place of no suffering and no doubt partly because for Dostoevsky's underground man "the crystal edifice [...] is forever indestructible" (Dostoevsky 2009, 33). Crystal is hard, adamantine, and reflective, a symbol of certitude, the sublime and sublimation.

Following on from Dostoevsky, the Crystal Palace for Berman (1988, 235, 237) is both "fact and symbol" and "the most visionary and adventurous building of the whole nineteenth century. Only the Brooklyn Bridge and the Eiffel Tower, a generation later, will match its lyrical expression of the potentialities of an industrial age." Yet rather than Dostoevsky, for Berman (1988, 239) "the most interesting and penetrating [contemporary] account of the Crystal Palace–the real one, that is, was written by [...] Lothar Bucher" (cited by Berman 1988, 239–240) whose account concludes with a rapturous hymn to "the transept which dissolves into a distant background where all materiality is blended into the atmosphere." The Crystal Palace, as with other exhibition buildings, is a capitalist secular cathedral to, and of, modernity and industry.[1]

Berman (1988, 240) notes that Bucher managed "to appropriate one of Marx's [and Engels'] richest images and ideas, 'all that is solid melts [or is sublimed] into air.' Like Marx, Bucher sees the tendency of solid material to decompose and melt as the basic fact of modern life." Yet this image and idea is precisely about both the sublime in which solid matter is transformed into a gas and as the basic fact of modern life. Decomposition occurs when solid matter breaks down into its constituent parts, not when

[1]Such as the Royal Exhibition Building in Melbourne with its nave and transept; see Giblett (2020, chapter 11). For sports stadiums as post-Christian hypercapitalist cathedrals, see Giblett (2020, chapter 10) and Sloterdijk (2014, 918; 2016, 573–602).

it is sublimed into the atmosphere. Melting occurs when a frozen liquid thaws back into its liquid state, not when solid matter is transformed into a gas. The sublime is the basic fact of modern life in which the solidities of traditional ways of life, cultures, and modes of production and consumption, etc. are transformed into thin air. Around the name of the sublime, as Lyotard (1989, 199) said, modernity triumphed, not least over wetlands. The slime of swamps, in which the solid earth is desublimated into liquid water, is the counter to the sublime of the city in which the solid earth is sublimated into gaseous air. The sublime city fills or drains the slimy swamp in order to create solid earth that it then sublimates into the gaseous heights of its ethereal structures like the Crystal Palace. The slimy swamp precedes and resists the sublime city. The sublime city reverts to feral slimy swamp in major floods.

More recently, the Crystal Palace for Sloterdijk is an architectural archetype, or archetypal architecture, for modernity. Indeed, for Sloterdijk it is a more fitting and apt architectural type of modernity than the arcades that made Paris for Benjamin "the capital of the nineteenth century" (Sloterdijk 2013, 170, 173–176; 2016, 323; Benjamin 1999a). London for Sloterdijk (2016, 323) was "the capital of the nineteenth century, even before Paris." After Sloterdijk (2013, 169) thinks of Dostoevsky"s "reference to Western civilization as a 'crystal palace'," he goes on to argue that "with its construction, the principle of the interior overstepped a critical boundary" as:

> it began to endow the outside world as a whole with a magical immanence transfigured by luxury and cosmopolitanism. Once it [the world] had been converted into a large hothouse and an imperial culture museum, it revealed the timely tendency to make both culture and nature indoor affairs. (Sloterdijk 2013, 170; see also 286 n1; 2016, 162–164; 2018, 142)

And to make the world an indoor affair inside the hothouse of 'global warming' or 'climate change' in the Anthropobscene. The Crystal Palace represents for Sloterdijk (2013, 171)"comfort and convenience" and "self-fulfilment of the consumer." It also represents discomfort, inconvenience and self-diminution for all beings (including the consumer) in the carboniferous capitalist hothouse with hotter days and hotter bush and forest fires burning more frequently around the world contrasted with the green house of the swamp world in which London and other cities were founded.

References

Ackroyd, Peter. 2001. *London: The Biography*. New York: Random House.

Ackroyd, Peter. 2011. *London Under*. London: Chatto & Windus.

Benjamin, Walter. 1973a. *Charles Baudelaire: A Lyric Poet in the Era of High Capitalism*. Translated by Harry Zohn. London: Verso.

Benjamin, Walter. 1973b. *Illuminations*. Translated by Harry Zohn. London: Fontana.

Benjamin, Walter. 1979. *One-Way Street and Other Writings*. Translated by Edmund Jephcott and Kingsley Shorter. London: NLB.

Benjamin, Walter. 1999a. *The Arcades Project*. Translated by Howard Eiland. Cambridge, MA: Belknap Press of Harvard University Press.

Benjamin, Walter. 1999b. *Selected Writings: Volume 2, 1927-1934*. Edited by Michael Jennings, Howard Eiland and Gary Smith. Cambridge, MA: Belknap Press of Harvard University Press.

Berman, Marshall. 1988. *All That Is Solid Melts into Air: The Experience of Modernity*. New York: Penguin.

Brant, Clare. 2007. Artless and Artful: John Gay's *Trivia*. In *Walking the Streets of Eighteenth-Century London: John Gay's Trivia (1716)*, ed. Clare Brant and Susan E. Whyman, 105–119. Oxford: Oxford University Press.

Brown, Norman O. 1959. *Life Against Death: The Psychoanalytical Meaning of History*. Hanover, NH: Wesleyan University Press.

Coolidge, John. 1994. *Gustave Doré's London: A Study of the City in the Age of Confidence 1848-1873*. Dublin, NH: William L. Bauhan.

Davis, Mike. 2006. *Planet of Slums*. London: Verso.

Dostoevsky, Fyodor. 2009. *Notes from Underground*. Translated by Boris Jakim. Grand Rapids, MI: William B. Eerdmans. First published in Russian in 1864.

Doyle, Arthur Conan. 1981. *The Penguin Complete Sherlock Holmes*. London: Allen Lane.

Engels, Friedrich. 1987. *The Condition of the Working Class in England*. Edited by Victor Kiernan. London: Penguin. First published in German in 1845.

Evaristo, Bernardine. 2002. *The Emperor's Babe: A Novel*. London: Penguin.

Fussell, Paul. 2000. *The Great War and Modern Memory*. New York: Oxford University Press. First published 1975.

Galeano, Eduardo. 2009. *Mirrors: Stories of Almost Everyone*. Translated by Mark Fried. New York: Nation Books.

Gay, John. 2007. Trivia: Or, the Art of Walking the Streets of London. In *Walking the Streets of Eighteenth-Century London: John Gay's Trivia (1716)*, ed. Clare Brant and S. E. Whyman, 167–215. Oxford: Oxford University Press. First published 1716.

Giblett, Rod. 2020. *Modern Melbourne: City and Site of Nature and Culture*. Bristol: Intellect Books.

Gissing, George. 1992. *The Nether World*. Oxford: Oxford University Press. First published 1889.

Hall, Peter. 2014. *Cities of Tomorrow: An Intellectual History of Urban Planning and Design Since 1880*. 4th ed. Chichester: Wiley Blackwell.

Hunt, Margaret R. 2007. The Walker Beset: Gender in the Early Eighteenth-Century City. In *Walking the Streets of Eighteenth-Century London: John Gay's Trivia (1716)*, ed. Clare Brant and Susan E. Whyman, 120–130. Oxford: Oxford University Press.

Jenner, Mark. 2007. 'Nauceious and Abominable'?: Pollution, Plague, and Poetics in John Gay's *Trivia*. In *Walking the Streets of Eighteenth-Century London: John Gay's Trivia (1716)*, ed. Clare Brant and Susan E. Whyman, 90–104. Oxford: Oxford University Press.

Jerrold, Blanchard, and Gustave Doré. 2005. *London: A Pilgrimage*. London: Anthem Press. (First published 1872).

Kotkin, Joel. 2005. *The City: A Global History*. New York: Random House.

Landa, Manuel de. 2000. *A Thousand Years of Nonlinear History*. New York: Swerve Editions.

Lehan, Richard. 1998. *The City in Literature: An Intellectual and Cultural History*. Berkeley, CA: University of California Press.

London, Jack. 2001. *The People of the Abyss*. London: Pluto Press. First published 1903.

Long, David. 2014. *A History of London in 100 Places*. London: Oneworld.

Lyotard, Jean-François. 1989. The Sublime and the Avant-Garde. In *The Lyotard Reader*, ed. Andrew Benjamin, 196–211. Oxford: Basil Blackwell.

Mayhew, Henry. 1985. *London Labour and the London Poor*. London: Penguin. First published 1861–62.

Melosi, M. V. 2008. *The Sanitary City: Environmental Services in Urban America from Colonial Times to the Present*. Pittsburgh, PA: University of Pittsburgh Press.

Morrison, Arthur. 1996. *A Child of the Jago*. Chicago: Academy. First published 1896.

Mumford, Lewis. 1961. *The City in History: Its Origins, Its Transformations, and Its Prospects*. New York: Harcourt.

Neuwirth, Robert. 2006. *Shadow Cities: A Billion Squatters, a New Urban World*. New York: Routledge.

Picon, Antoine. 2000. Anxious Landscapes: From the Ruin to Rust. *Grey Room* 1, 65-83. Accessed September 29, 2006. https://www.gsd.harvard.edu/wp-content/uploads/2016/06/picon-ruinerouille.pdf

Sloterdijk, Peter. 2012. The Time of the Crime of the Monstrous: On the Philosophical Justification of the Artificial. In *Sloterdijk Now*, ed. Stuart Elden, 165–181. Translated by Wieland Hoban. Cambridge, UK: Polity Press.

Sloterdijk, Peter. 2013. *In the World Interior of Capital: For a Philosophical Theory of Globalization*. Translated by Wieland Hoban. Cambridge, UK: Polity Press. First published in German in 2005.
Sloterdijk, Peter. 2014. *Spheres, Volume 2: Globes: Macrospherology*. Translated by Wieland Hoban. Los Angeles, CA: Semiotext(e). First published in German in 1999.
Sloterdijk, Peter. 2016. *Spheres, Volume 3: Foams: Plural Spherology*. Translated by Wieland Hoban. Los Angeles, CA: Semiotext(e). First published in German in 2004.
Sloterdijk, Peter. 2018. *What Happened in the 20th Century*. Translated by Christopher Turner. Cambridge, UK: Polity Press. First published in German in 2016.
Smith, P. D. 2012. *City: A Guidebook for the Urban Age*. London: Bloomsbury.
Thomas, Edward. 1978. Diary of Edward Thomas [1917]. In *The Collected Poems of Edward Thomas*, ed. R. George Thomas, 460–481. Oxford: Clarendon.
Thomson, James. 1993. *The City of Dreadful Night*. Edinburgh: Canongate Classics. First published 1880.
Thoreau, Henry David. 1962. *The Journal of Henry D. Thoreau*. Vols. I–XIV. Edited by Bradford Torrey and Francis Allen. New York: Dover.
Virilio, Paul and Sylvere Lotringer. 1983. *Pure War*. Translated by M. Polizzotti, New York: Semiotext(e).

CHAPTER 10

Tombs, Wombs, and Being Born

Abstract The Greek philosophy of the four elements was syncretised with Judeo-Christian theology in the seventeenth century CE by John Milton in *Paradise Lost* in which he rewrote the biblical story of creation out of what he called "the womb of Nature," the pre-mixing of the four elements. In imperialist adventure-romances the journey of the hero into the swamp figures the horror of repeating birth in reverse of going up the birth canal. J. G. Ballard's post-apocalyptic dystopian novel, *The Drowned World*, is concerned with the return of the spatial repressed (the tropical colonies and the uterus), temporal repressed (the uterine and colonial pasts) and psychological repressed (memories of the womb and the mother's body). In Janet Frame's story, the lagoon represents a way for the first person narrator of establishing and celebrating matrilineal descent. Lagoons are some of the most fertile ecosystems and habitats on the planet. Yet the lagoon at low tide is not a place of new life, but of death and destruction, part of the rhythm of life and death. Lagoons and other wetlands are places of what Hannah Arendt called natality, or being born, that she countered to Martin Heidegger's morbid solipsism of being towards death.

Wetlands are wombs of new life and tombs of old life out of which new life springs. In the seventeenth century John Milton rewrote the biblical story of 'Paradise Lost' and its backstory beginning with creation. For him,

R. Giblett, *Living with Nature in the Anthropobscene*,
https://doi.org/10.1007/978-3-032-12822-5_10

creation came out of what he called "the womb of Nature," the premixing of the four elements, an apt descriptor for wetlands. In imperialist adventure-romances, such as Rider Haggard's *She* and Joseph Conrad's 'The Lagoon,' the journey of the hero into the swamp figures the horror of repeating birth in reverse. In J. G. Ballard's post-apocalyptic dystopian novel, *The Drowned World*, the world has drowned because of solar flares, but it is applicable to rising sea levels in the age of global boiling. It is concerned with the return of the spatial repressed (the tropical colonies and the uterus), temporal repressed (the uterine and colonial pasts) and psychological repressed (memories of the womb and the mother's body). By a curious literary historical irony, in Janet Frame's story of the same title as Conrad's the lagoon represents a way for the first person narrator of establishing and celebrating matrilineal descent. The lagoon at low tide is not a place of new life, but of death and destruction. It is a part of the rhythms of life and death. Lagoons and other wetlands of places of what Hannah Arendt called natality, or being born.

10.1 Nature's Womb and Tomb

John Milton, the seventeenth-century syncretist of Judeo-Christian theology and Greek philosophy, followed Aristotle by only admitting the possibility of the mixture of four elementary qualities into what he called "the cumbrous elements, earth, flood, air, fire" (III, 715):

> [...] ye elements, the eldest birth
> Of Nature's womb, that in quaternion run
> Perpetual circle, multiform, and mix
> And nourish all things, let your ceaseless change
> Vary to our great Maker still new praise. (V, 180–184)

The mixing of qualities in "Nature's womb" to procreate the elements does indeed nourish all things as Milton indicates, but not only the four mixtures of qualities into the four elements that Aristotle and Milton permit. Indeed, the two aberrant mixtures of wet and dry, hot and cold produce the temperate and tropical wetland which are far more productive than both dry land and cold water. It is the wet land and warm water which give birth to life and which are the womb of the world.

Rather than the womb of the world, wetlands for Milton are the tomb in the underworld. In *Paradise Lost* Milton saw the Styx, one of rivers of the classical underworld and for Dante a slimy swamp in the fifth circle of the inferno (as we saw in Chap. 3), as "the burning lake," as "abhorrèd" and as "the flood of deadly hate" (II, 576, 577). In the same book of *Paradise Lost* he elaborates later in more detail on:

> [...] lakes, fens, bogs, dens and shades of death,
> A universe of death, which God by curse
> Created evil, for evil only good,
> Where all life dies, death lives, and Nature breeds,
> Perverse, all monstrous, all prodigious things,
> Abominable, inutterable, and worse
> Than fables yet have feigned, or fear conceived,
> Gorgons and Hydras, and Chimeras dire. (II, 621–628)

Later in the same book of *Paradise Lost* Milton describes in more detail the watery world and womb of the first day of creation that morphs into Hell:

> [...] this wild abyss,
> The womb of Nature and perhaps her grave,
> Of neither sea, nor shore, nor air, nor fire,
> But all these in their pregnant causes mixed
> Confus'dly, and which thus must ever fight,
> Unless th' Almighty Maker them ordain
> His dark materials to create more worlds,
> Into this wild abyss the wary Fiend
> Stood on the brink of Hell. (II, 910–918)

Milton's Satan, referred to here as "the wary Fiend," is a swamp serpent and marsh monster, or more precisely a miasmatic marsh monster, an elemental aberration who mixes earth, air, fire and water (as we saw in Chap. 2). The womb of nature also gives birth to Satan who is cocreated into this wild abyss on the brink of Hell.

For Milton the womb (and perhaps the tomb) of nature is neither earth, water, air, nor fire. It mixes the elements in gestating and giving birth to life. This is an apt descriptor for a wetland that mixes solid earth, liquid water, gaseous air and hot fire (especially in the tropics). In the beginning of creation for Milton, everything begins in struggle, or war. Milton takes the pre-Socratic Heraclitus' view that war is "father/ of all things and, king" (Heraclitus 2003, 29; see Kirk et al. 1983, 195–196).

Milton's God creates this world and others from "His dark materials," a term purloined by Phillip Pullman for the title of his fantasy trilogy.

God, for Milton in keeping with Genesis 1:2, brings order out of the wetland chaos, out of "the vast immeasurable abyss/ Outrageous as a sea, dark, wasteful, wild" (VII, 211,2):

> [...] on the wat'ry calm
> His brooding wings the Spirit of God outspread,
> And vital virtue infused, and vital warmth
> Throughout the fluid mass, but downward purged
> The black tartareous cold infernal dregs
> Adverse to life. (VII, 234–239)

Milton also takes Heraclitus's view that fire in the sense of energy is the archetypal, formative matter flowing through all things (Heraclitus 2003, 15–17; Kirk et al. 1983, 199–200). He also takes the Platonic/Socratic view that divides the heavens from the dregs, valorises the heavens and denigrates the dregs (as we saw in Chap. 2). He syncretises these views from Geek philosophy with Judeo-Christian theology.[1]

Hell as swamp was not just any part of hell for Milton, but the lowest part of the underworld. Milton reproduces part of what has been called "the naive view of the world" in which Tartarus was "the lowest part of the underworld" (Kirk et al. 1983, 9). In *The Iliad* "misty Tartaros" is "the deepest gulf below earth" and in Hesiod's *Theogony* it is "the grievous, dank limits which even the gods detest" (Kirk et al. 1983, 9 and 40). Milton regards water as lifeless until the spirit (or breath as they are the same word in Hebrew and Greek) of God breathes life into it. He rejects both "Thales' idea[s] of the earth floating on water" and that water is "in some way the source of all things" (Kirk et al. 1983, 11 and 88). Water is the mother of all things, and goddess.

Milton's theology entails sublimation of the materiality and immanent spirituality of earth and water into the transcendental spirituality of God. For Milton the Spirit of God replaces water in Thales's pre-Socratic cosmogony. Whereas in the latter, "water is the continuing, hidden constituent of all things" (Kirk et al. 1983, 94), for the former the Spirit of God

[1] William Blake critiqued that "war is energy enslaved" (as we saw in Chap. 7). In Heraclitean terms, this means that war enslaves the creative and destructive power of fire and uses it for destructive ends alone (as we saw especially in Chap. 8). War also recreates, in Heraclitean terms, unbalanced strife.

becomes this secret, life-giving essence. Whereas in the latter, "all living things depend on water for nourishment" (94), for the former all living beings depend on the Spirit of God for life. Whereas the latter regard "the nurture of all things to be moist" (89), for the former the nurture of all things is dry and windy. Milton's position is not only indicative of a sublimation away from the materiality and immanent spirituality of water to the transcendental spirituality of God, but also of a fundamental shift in the conceptualisation of human life from being constituted by fluids to being indicated by breath (and Spirit), and of the human body from being regarded as a system of fluids, or wetland, to a bag of winds, of hot air.

Milton's theology has little or no biblical basis or warrant in this regard, but comes from Western philosophical idealism that separates spirit and matter. It also valorises spirit as life-giving force over matter as lifeless substance until spirit infuses it with life (as Milton says). This kind of syncretism also characterises ecotheology, such as the distinction between spirit and matter not found in the Bible. Even though breath and spirit are the same word in Hebrew (and Greek), Theodore Hiebert (2008, 18) argues that "it is improbable that the word 'spirit' is ever a viable interpretation for the biblical Hebrew רוּחַ." He translates רוּחַ as breath, air, the atmosphere, God's breath and "the first sacred thing"—and not as spirit. He adds that to interpret the biblical Hebrew רוּחַ as spirit "can only seriously distort the biblical perspective" as the Hebrew Bible does not separate out spirituality and materiality, but sees them as interconnected. Hiebert argues that the "post-biblical theological dualism" of matter and spirit comes from Western philosophical idealism originating with Plato (as we saw in the first chapter). In addition, the concept of matter itself comes from western philosophy more generally, and matter as made up of indivisible particles comes from Western philosophical atomism originating with Democritus (Lloyd and Sivin 2002, 150).

As Luce Irigaray (1991, 13) has remarked in her imaginary dialogue with Fredrich Nietzsche, "it is always hot, dry and hard in your world," as it is in Milton's and as it is in the Stoics' for whom, as Cicero (1972, 134) puts it "all the elements of the universe are sustained by heat [...] it provides the life-force and is the source of all that comes to be." In the wetland world of the mother, it is always cool, wet and soft. The wetland, like the forest for John Evelyn, is "the moist womb of the world" (cited by Thoreau 1993, 66). The wetland provides the life-force and is the source of all that comes to be.

For Milton, like Nietzsche as Irigaray (1991, 7) also has remarked, "holes mean only the abyss for you," hence the need to invaginate and penetrate surfaces as a way of denying and repressing the depths of the Body of the Mother. Traditionally the extension of the swamp has been invaginated so it can be penetrated by phallic heroes (as we will see in the present chapter). A journey across or through a swamp becomes a journey penetrating an invaginated surface as a way of denying or repressing the depths of the Body of the Mother. In patriarchy, for Lyotard (1993, 21), "there are no holes, only invaginations of surfaces" as a way of denying and repressing the holes and depths of the mother's body.[2]

Typically in mythology, for Erich Neumann (1955, 157), "the underworld, the earth womb [...] is experienced in the archetypal nocturnal sea voyage of the sun or the hero." Typically in the epic, the hero descends into the underworld of the earth womb to meet and slay monsters that were born and live there, and re-emerge with his manhood intact and validated. Sloterdijk (2024, 6) argues that in the hero "the adventure of civilization begins as the colonization of egoic solid ground – the inhabiting and thronging of a new continent: autonomy, power, will, and knowledge." Before that, though, the hero has to colonise the uncanny slimy swamp of the id and the dark continent of Africa and female sexuality. Typically in post-epic narratives, especially in imperialist adventure-romances set in Africa, such as Rider Haggard's *She* and Joseph Conrad's 'The Lagoon,' the underworld, the earth womb, is experienced as an archetypal swamp journey of the hero.

A journey into the swamp goes up the birth canal which leads to the womb where life is recreated. Sandra Gilbert and Susan Gubar (1989, 13) argue that in Haggard's *She*:

> as the men make their way inland, through vaporous marshes and stagnant canals, the [wet]landscape across [?through] which they journey seems increasingly like a Freudian female *paysage moralisé* ['moralised landscape' or more precisely moralised wetlandscape or *marécage moralisé*] [...] and, in a symbolic return to the womb, they are carried up ancient birth canals through a rocky defile into 'a vast cup of earth' that is ruled by *She-who-must-be-obeyed*.

[2] For Lyotard (1993, 250), "the theoretical is a major procedure of invagination." The theoretical sublime invaginates the surface of the earth and interiorises its depths in an inversion of its spatial arrangement. Perhaps Fig. 3.1 in Chap. 3 should be turned upside down!

This cup is "'a spot where the vital forces of the world visibly exist'," "the very womb of the world" (Haggard 1991, 29 and 279).

The destination of the heroic journey up the birth-canal is for Neumann (1955, 149) "the womb of the earth." In patriarchal and filiarchal culture, the source of life is transformed into an orally sadistic hell, into what Neumann (1955, 149) calls "the deadly devouring maw of the underworld [...] the abyss of hell, the dark hole of the depths, the devouring womb of the grave and of death, of darkness without life, of nothingness" in a blatant and disavowing fantasy of reversal and negation. Womb becomes tomb in what Alice Jardine (1985, 32; her emphasis) calls "*the* founding fantasy: the active negation of the Mother" based on and repeating *the* founding act of what Irigaray (1993, 11) calls "the murder of the woman-mother," and what Sloterdijk (2025, 178–179) calls "matricide." This fantasy inverts the idea of tomb as womb for the birth of new life into womb as tomb for interring old life, inverting the idea that death is not the end of life as such, but the beginning of new life into the idea that new life is a kind of death.[3] Sloterdijk (2024, 102) argues that "whoever assumes the womb is tomb interprets the will to return into the originary chamber as a malign regression." Whoever assumes the tomb is womb interprets the will to return into the originary chamber as a benign progression. There is a hell of a lot of difference culturally and politically between the tomb as womb with its promise of new life and the womb as tomb with its threat of death. In this fantasy not only is, as Gilbert and Gubar (1989, 17) put it, "the womb of the Great Mother [...] also a tomb," but the swamp is transformed from the womb of life into the jaws of death.

Similarly in Joseph Conrad's first published story, 'The Lagoon,' the anonymous white man travels through black water up a narrow creek which is described as being:

> like a ditch: tortuous, fabulously deep; filled with gloom under the thin strip of pure and shining blue of the heaven. Immense trees soared up, invisible behind the festooned draperies of creepers. Here and there, near the glistening blackness of the water, a twisted root of some tall tree showed amongst the tracery of small ferns, black and dull, writhing and motionless like an arrested snake. The short words of the paddlers reverberated loudly

[3] For the tomb as womb see Marija Gimbutas (1989, 151–159). See also Monica Sjöö and Barbara Mor (1991, 8, 46, 51, 52, 59, 76, 103, 113, 180 and 184).

> between the thick and sombre walls of vegetation. Darkness oozed out from between the trees, through the tangled maze of the creepers, from behind the great fantastic and unstirring leaves; the darkness mysterious and invincible; darkness scented and poisonous of impenetrable forests. The men poled in the shoaling water. The creek broadened, opening out into a wide sweep of a stagnant lagoon. The forests receded from the marshy bank. (Conrad 1977, 172)

The passage through the swamp, the tunnel beneath its canopy, has often been used as a figure in patriarchal western culture, as David Miller (1989, 25) puts it, for "the passageway to the womb." A journey into the canal or tunnel of the swamp figures for patriarchy the horror of repeating birth in reverse. On these horrific journeys phylogenesis and ontogenesis are repeated in reverse by the patriarchal hero. Phylogenesis in reverse, or going back into the primeval slime of the swamp, repeats ontogenesis in reverse, going back into the black waters of the womb.

10.2 The Drowned World

Both Richard Jefferies's *After London* and J. G. Ballard's *The Drowned World* are post-apocalyptic dystopian novels concerned with the return of the spatial repressed (the tropical colonies and the uterus), temporal repressed (the uterine and colonial pasts) and psychological repressed (memories of the womb and the mother's body). Indeed, for *The Drowned World* there is a homology between all three. Unlike Jefferies's vision in which London is flooded as a return of the drained marshes of the Thames, in *The Drowned World* the previously drained world, the temperate dry lands, of London is flooded as a return of the repressed and colonised tropical 'other,' the tropical wet lands. In *After London* a return of the repressed within, the endocolonised marshes of temperate England, takes place; in *The Drowned World*, a return of the repressed without, the exo-colonised lagoons of the tropics, occurs. The boundary between the tropical and the temperate ceases to exist in *The Drowned World* with the former's region and climate having taken over the latter's. London, and the whole of England for that matter, is now located in "the European lagoons" (Ballard 1983, 12).

The inhabitants of London are now so many latter day colonists of the tropical 'at home.' The lagoon which was not 'at home,' but was colonised in the tropical, has come home; the wetlands which were not like those 'at home' have come home and made home their own; the unhomely has invaded and conquered the homely. In this return of the colonial repressed the official job of Kerans, the central character of the novel, is to serve a vestigial colonial function of cartographer by "mapping the shifting keys and harbours and evacuating the last inhabitants [...] living on in the sinking cities" (12). As the story progresses, however, Kerans is transformed from imperialist cartographer to wetlands conservationist. Wetlands are difficult to map spatially because of their changing status temporally. By an etymological play on 'lagoon' and 'lacuna,' both of which come from the Latin '*lacuna*' meaning pool, Kerans's job is made more difficult because a lacuna (and a lagoon on a map) is "a hiatus, blank, missing part, a gap, an empty space, spot, or cavity," like a cloaca.[4] For the cartographer the lagoon, the slime (and the wetland more generally) is an impossible abject because of its changing shape.

Yet Kerans undergoes a transformation in his perception of the wet land. Initially, for him "the lagoon was nothing more than a garbage-filled swamp" (Ballard 1983, 13) like Jefferies's London. Culturally, and spatially, the two places are not so far apart as wetlands have often been the site of 'sanitary land fill.'[5] The description of the swamp as "nothing more [...]" upholds the western tradition of reductive, pejorative markers being assigned to wetlands. Riggs concurs with Kerans's summation when he pronounces that "the whole place is nothing but a confounded zoo" (17), which is like Jefferies's reduction of the city to the animal. Yet unlike Jefferies's swamp, which is unhealthy in accordance with the prevailing miasmatic theory of disease of his time, Ballard's lagoon simply smells bad: "Kerans felt the terrible stench of the water-line, the sweet compacted smells of dead vegetation and rotting animal carcasses" (13). Kerans feels, rather than smells, the stench. Perhaps he would rather feel the stench than smell it. The sensation of touch stays on the outside, whereas the sensation of smell penetrates inside. For Kerans "a thick cloacal stench exuded from the silt flat" (61).

The returns of the repressed, of the bad smells of the swamp, of the colonial other, also involve the return to the past, both the past of the human species and the individual past, in a reversal of the Freudian tenet

[4] See *The Shorter Oxford English Dictionary*, 1167.

[5] For a graphic illustration see the photographs by Joseph Muench in Niering (1966, 164, 165) and by Wayne Lawler in McComb and Lake (1990, 225).

following Ernst Haeckel that individual development repeats the development of the species (ontogenesis repeats phylogenesis). In *The Drowned World*, the apocalypse of the world drowning reverses phylogenesis in a return to the primeval slime, or more precisely to "impenetrable Mato Grossos sometimes three hundred feet high [...] a nightmare world of competing organic forms returning rapidly to their Paleozoic past" (19).

Ontogenesis is also reversed for Kerans in a return to amniotic fluid, a prospect which fills him initially with horror when he wonders whether "'perhaps these sunken lagoons simply remind me of the drowned world of my uterine childhood - if so, the best thing is to leave straight away'" (28). The lagoon, perhaps of all wetlands, is markedly, though naturally, enclosed. On a continuum of wetland types from the most earthbound to the least, the lagoon is the most wet of the wetlands. It is also separated or closed off from the sea by low sandbanks or by an atoll, a small sea within a larger. The lagoon is a kind of natural vessel made, like all wetlands, of earth and water, but with a thinner shell, a rim and even a lip. The drowned world is referred to as "the black bowl of the lagoon" (47, see also 70). No doubt the black bowl contains black water.

This bowl of black water is archetypally the womb. In his archetypal study of the Great Mother, Neumann (1955, 47) has suggested that:

> the natural elements that are essentially connected with vessel symbolism include both earth and water. This containing water is the primordial womb of life, from which in innumerable myths, life is born. It is the water 'below,' the water of the depths, ground water and ocean, lake and pond.

And lagoon. These waters are highly fertile. In particular, the littoral, the intertidal zone, the estuarine and lagunine, are some of the most productive ecosystems on the planet, up to twenty times richer in wildlife than the open ocean (Earll 1991, 161).

For Kerans the drowned worlds of the phylogenetic and ontogenetic, of the exterior and interior, past and present become indistinguishable. The uterine past of the individual and the paleozoic past of the human species have been repressed. Their joint return are represented in the iconography of the drowned world itself and portrayed in "one of Max Ernst's self-devouring phantasmagoric jungles [which] screamed silently to itself, like the sump of some insane unconscious" (Ballard 1983, 29). The unconscious is even more strongly figured in aquatic terms in the reference to "the submerged levels below his consciousness" (83). *The*

Drowned World seems to have as equally a strong intertextual relationship with Max Ernst's painting, 'Europe After the Rain,' with its nightmarish vision of what Robert Hughes (1980, 255) calls "a panorama of a fungoid landscape seen as though in the aftermath of an annihilating biblical deluge" as it does with *After London* with its apocalyptic vision of the city as a stagnant swamp of dead black water after the flood.

A powerful gender politics and psychodynamics is enacted here between the swamp, the unconscious and the female. The swamp has been used a figure for the unconscious and for the female, especially female sexuality, in patriarchy. Neumann (1955, 260) argues that:

> the female powers dwell not only in ponds, springs, streams and swamps but also in the earth, in mountains, hills, cliffs, and - along with the dead and unborn - in the underworld. And above all, the mixture of the elements of water and earth is primordially feminine; it is the swamp, the fertile muck, in whose uroboric nature the water may equally well be experienced as male and engendering or as female and birthgiving.

The lagoon is the most watery mixture of earth and water, the wetland with the least possible amount of land, and so perhaps the wetland to be most strongly feminised in the western tradition.

To complete the triangle of troping between the unconscious, the feminine and the swamp, patriarchy has gendered the unconscious as feminine and the conscious as masculine (see Neumann 1955, 147–148). Luce Irigaray (1985b, 73) has wondered in a critique of the power of discourse "whether certain properties attributed to the unconscious may not, in part, be ascribed to the female sex, which is censured by the logic of consciousness. Whether the feminine has an unconscious or whether it is the unconscious."[6] One of the properties attributed to the unconscious which has also been ascribed to the female sex is fluidity. Indeed, Irigaray (1985b, 116) maintains in an essay on the mechanics of fluid that "historically the properties of fluids have been abandoned to the feminine."[7] Yet it is only certain properties of certain fluids which have been abandoned to the feminine, primarily the static properties of stagnant fluids. The dynamic properties of flowing fluids have been appropriated by the masculine.[8]

[6] See also the interview with her (Irigaray 1977, 70) and Irigaray (1985a, 111 and 224).

[7] Similarly, though essentialistically and ahistorically, for Camille Paglia (1991, 91) "female experience is submerged in the world of fluids."

[8] Neumann (1955, 48) suggests something along similar lines.

Rivers have been masculinised and swamps feminised, the latter most powerfully in the organising principle of *The Drowned World*, the wetland as womb, though the novel develops this trope from the modality of horror to that of fascination. Rather than enacting a typically masculine repudiation of his birth, Kerans becomes fascinated with his uterine experience, though not with his mother per se. He shows no interest in his mother as a person. His fascination with his uterine experience is abstracted from the body in which it occurred. Kerans's mother is conspicuous by her absence from his memory and his curiosity. She is a hiatus, blank, missing part, a gap, an empty space, in short, a lacuna in the story of his life, a lagoon in the map of his life.

Perhaps not surprisingly then, his attitude to Beatrice, the only female character in the novel, is inflected with misogyny. The choice of name is rather ironic given Dante's fascination with another Beatrice and with the slimy in Canto 7 of the *Inferno*. The wetland hell of *The Drowned World* could be read as a postmodern version of Dante's medieval vision of a Stygian marsh with its souls of the sullen stuck in slime. For Kerans, though, the Beatrice of *The Drowned World* is not the beatific Beatrice of Dante's *Divine Comedy*, but "Pandora with her killing mouth and witch's box of desires and frustrations, unpredictably opening and shutting the lid" (Ballard 1983, 31). Beatrice is figured as an automata, that ultimate figure of the *unheimlich* or uncanny for Freud, with a devouring, mechanical and oral-sadistic mouth.

In an essentialist gesture, for Neumann (1955, 122–123), "the mouth as rending, devouring symbol of aggression is characteristic of the dangerous negative elementary character of the Feminine." Or more precisely, the fear of being eaten by the mother in an act of symmetrical reciprocity characterises the oral-sadistic phase interposed between what Freud (1984, 419) calls in the essay 'The Economic Problem of Masochism' the "primitive oral organisation - the fear of being eaten up [...] by the father" and the "anal-sadistic phase - the wish to be eaten by the father." The fear of symmetry and fairness being exercised back against oneself characterises the oral-sadistic.

As an enclosed, moist space with a single opening the mouth, especially before teeth appear, is associated with the womb in the anal phase where both are seen as passive organs. For Marie Bonaparte (1953, 24) "in the earlier anal stage, it was the cloaca that appeared as the erotogenic organ. Then, as throughout life, the cloaca is a passive organ as largely is the mouth before the teeth appear, in the primary stage." But when teeth appear the mouth, and the cloaca by previous association, can become active, even sadistic. The functions of both are inverted with the mouth becoming the organ of (re)production as the cloaca was (and still is) and the cloaca the site of consumption and destruction as the mouth had been.

For Neumann (1955, 168):

> the positive femininity of the womb appears as a mouth; that is why 'lips' are attributed to the female genitals, and on the basis of this positive symbolic equation the mouth, as 'upper womb,' is the birthplace of the breath and the word, the Logos. Similarly, the destructive side of the Feminine, the destructive and deadly womb, appears most frequently in the archetype form of a mouth bristling with teeth [...] the vagina dentata.[9]

Beatrice's "killing mouth" and Pandora's box is a vagina dentata for Kerans and Kerans's fascination with his uterine past is displaced onto and re-enacted with her as a fear of ingestion and entrapment.

For whom or for what, though, is the womb, in Neumann's terms, destructive and deathly? Only for the phallic man and the phallus, as Neumann (1955, 171) goes on to suggest, albeit in misogynist terms: "the Feminine - as avid womb - attracts the male and kills the phallus within itself in order to achieve satisfaction and fecundation." Yet by phantasising the vagina dentata as an orally sadistic organ, the phallic man is in fact an impotent man as Bonaparte (1953, 24) suggests that "the cloaca [...] will never be dentated except in the phantasies of impotent men who fear it, as such, in woman." For many men the womb, Irigaray (1993, 16) argues, "is variously phantasized as a devouring mouth, as a sewer in which anal and urethral waste is poured, as a threat to the phallus or, at least, as a reproductive organ" as the wetland is phantasised as a devouring mouth, as a sewer in which anal and toxic waste is poured, as an engulfing threat to phallic power, or as reproductive organ.

Women's desire, Irigaray (1985b, 29–30) argues, is:

> often interpreted, and feared, as a sort of insatiable hunger, a voracity, that will swallow you whole. Whereas it really involves a different economy more than anything else, one that upsets the linearity of a project, undermines the goal-object of a desire, diffuses the polarisation toward a single pleasure, disconcerts fidelity to a single discourse.

[9] Irigaray in several places has propounded a far more positive valuation of the association between lips and female sexuality; see, for example Irigaray (1977, 65).

The fluidity of women's desire threatens to undermine the shores, banks, dykes, etc. of men's objects, private property, things, turds, etc., just as the sea and the swamp threaten to absorb 'developments' built between them.[10]

For Neumann (1955, 171) it is only "the hero" who can escape this fate of re-absorption:

> the womb, the gate, the gullet [...] actively swallows, devours, rends and kills. Its sucking power is mythologically symbolised by its lure and attraction for man, for life and consciousness and the individual male, who can evade it only if he is a hero, and even then not always.

For the phallic man the womb is an orally sadistic hell sucking at him, like the slime which sucks at Sartre, which he can only escape if he is prepared to descend heroically into it and find his way through it. In fact, heroism is constituted in patriarchy by the ability to survive the descent into the underworld, or 'dark continent,' or 'black water,' of female sexuality as I have shown in an earlier chapter. Kerans is a hero who does literally descend into the underworld of the drowned world, or womb, of London and not only survives but undergoes a transformation, or kind of rebirth.

Kerans's fascination with his uterine past and its homologous relationship with the Paleozoic past of the human race are confirmed by Bodkin: "just as psychoanalysis reconstructs the original traumatic situation in order to release the repressed material, so we are now being plunged back into the archaeopsychic past, uncovering the ancient taboos and drives that have been dormant for epochs" (Ballard 1983, 43–44). Plunged back, in other words, into the repressed world of primeval slime and the swamp as a kind of recapitulation of birth in reverse and in the process raising the unconscious to consciousness in a successful 'talking cure.'

For Freud (1973, 323), "the task of a psychoanalytic treatment [...] is to make conscious everything that is pathogenically unconscious [...] and to fill up all the gaps in the patient's memory, to remove his amnesias." In other words, to fill up, and so fill in, the lacunae (lagoons) in the surfaces

[10] Ruth Barcan, in a paper entitled 'Natural Histories: Nature, History and Gender on the Gold Coast' presented at the Australian Cultural Studies Conference in December 1991 argued that "the Gold Coast, built on the swamp at the edge of the sea, has never fully rid itself of the terror of being re-absorbed"—from both sides in the age of global boiling and rising sea levels. See also Irigaray (1985a, 237) for further on the fluid.

of the patient's memory about the womb/wetlands as the place from whence we came and as the underside of the city and thus to repeat colonising writing. The task of a psychoanalytic ecology, on the other hand, would be to read the absences in the maps of the wetlands/womb as symptomatic of a psychogeopathology of the will to fill wetlands, and so decolonise writing.

Bodkin goes on to suggest that "the uterine odyssey of the growing foetus recapitulates the entire evolutionary past" (Ballard 1983, 44), or in other words, ontogenesis recapitulates phylogenesis from primeval slime to hypermodern times as a heroic journey. Individually, and ontogenetically, for Bodkin, "'we really remember these swamps and lagoons'" (48). The lacunae in memory and the lagoon on the map are filled up. The lagunine apocalypse of a decaying Europe, however, recapitulates in reverse the entire evolutionary past from hypermodern times back to primeval slime to arrive at "'the very junction where we stand now on the shores of this lagoon, between the Paleozoic and Triassic Eras'" (44). This is both a form of time-travel in which the Earth (and its history) is traced temporally (which includes its wetlands unlike the spatial tracing of the Earth (and its geography) on maps which invariably do not trace wetlands) and of what Bodkin graces with the high-sounding name of "the Psychology of Total Equivalents" or "Neuronics":

> I am convinced that as we move back through geophysical time so we re-enter the amnionic [*sic*] corridor and move back through spinal and archaeopsychic time, recollecting in our unconscious minds the landscape of each epoch, each with a distinct geological terrain, its own unique flora and fauna, as recognisable to anyone else as they would be to a traveller in a Wellsian time-machine. Except that this is no scenic railway, but a total re-orientation of the personality. If we let these buried phantoms master us as they re-appear we'll be swept back helplessly in the flood-tide like pieces of flotsam. (Ballard 1983, 44–45)

Like Kerans, Bodkin fears being plunged or swept back in metaphors of fluidity into the past, whether it be the uterine or paleozoic past or both. It is matter of master or be mastered, or more precisely mistressed. Even though the tropical and the orient have won out climatically and geographically over the temperate and the occident, the battle still rages psychologically between these sites in the individual and collective mind and body. The colonies have struck back and triumphed externally, but internally the battle still goes on.

Much of that battle is fought out, and lost, as would be expected in accordance with Freud, in the domain of dreams. Kerans has a dream in which he steps out into the lake:

> whose waters now seemed an extension of his own bloodstream. As the dull pounding rose, he felt the barriers which divided his own cells from the surrounding medium dissolving, and he swam forwards, spreading outwards across the black thudding water. (Ballard 1983, 71)

This dream of the dissolution of Keran's ego in black water is a reversal of the dreams which symbolise the ego's formation. For Jacques Lacan in his essay, 'The Mirror Stage':

> the formation of the *I* is symbolised in dreams by a fortress, or a stadium - its inner arena and enclosure, surrounded by marshes and rubbish tips, dividing it into two opposed fields of contest where the subject flounders in quest of the lofty, remote inner castle whose forms [...] symbolise the id in a quite startling way. (Lacan 1977, 5)

Lacan bears out Sloterdijk's (2024, 4) conjecture when he asks rhetorically "was the psychoanalytic concept of the ego not invented ultimately to ban the uncanny to the outer edges of the autonomous life and to contest all its claims to a place in the center?" The uncanny is precisely not in the center, can never be admitted to the center and is always on the margins of the ego in the marshes and rubbish tips of the id. That is what the uncanny is.

The psychoanalytic concept of the ego was invented ultimately to marginalise the uncanny to the outer edges of the autonomous life and keep it at bay there. All attempts to ban the uncanny are as equally futile and unsuccessful as the attempts to ban other euphoric and soporific substances, such as illicit drugs. Did not the philosophic concept of the uncanny invite it into culture and give it a place at the table of the symposium of cultural creativity (as we have seen in a previous chapter)? Did not psychoanalysis inherit the dualist Cartesian *cogito* and its concept of the ego that drained and enclosed the commons of the wetland body into the private property of the individual mind, just as the enclosure acts and drainage did with wetlands, such as the Fens, in England?[11] Did not the psychoanalytic uptake of the concept of the uncanny acknowledge the virtual presence of the uncanny on the threshold between nature and culture? Freud not only invented the psychoanalytic concept of the ego, but also took up the uncanny and kept it active in the toolbox of cultural and critical theory (as we have seen too).[12] Sloterdijk (1987, 330) argued earlier that "the uncanny that was supposed to be banned comes to the fore again out of the means used to protect against it." The ego and the uncanny are not possible without each other, just as the sublime and the sublime city is not possible without slime, slimy swamps and slimy slums (as we have seen also). "Where ego is, there id was and shall still be, not unlike the draining of the English Fens."

The exchanges, as Miller (1989, 208) puts it, in the swamp, as well as in the lagoon (and other wetlands more generally), "between skin and enveloping matter [...] all initiate the dissolution of the self." In Kerans's dream he leaves the security of what Antony Easthope (1986, 35–44) calls, following in Kafka's footsteps, the castle of the self. Walter Benjamin (1973, 126; 1999, 806) writing on the tenth anniversary of Kafka's death argues that in Kafka's *The Castle*, "just as K lives in the village on Castle Hill, modern man lives in his body; the body slips away from him, is hostile towards him [...] Exile – his exile – has gained control over him [...] the most forgotten alien land is one's own body." This "most forgotten alien land" is more precisely the most forgotten alien wet-land of one's own body. Kerana flounders around in the black waters and garbage dump of London's (and Lacan's) lagoons. This dream is a recurring one which does not just involve the dissolution of the boundaries of his body and his ego, but also a return to his amniotic past. The lake becomes "the warm amnionic [*sic*] jelly through which he swam in his dreams" (Ballard 1983, 99). Yet this womb associated with life is also ambivalently, and misogynistically, figured as a tomb associated with death.

All Kerans's fears of ingestion and dissolution are re-enacted when, in the ninth chapter entitled, 'The Pool of Thanatos' (death-instinct), he dives into "a tank filled with warm, glutinous jelly that clamped itself to his calves and thighs like the foetid embrace of some gigantic protozoan monster" (104). Given that he is diving into water figured as amniotic fluid,

[11] For Descartes and the Cartesian *cogito* and its dualism, see Sloterdijk (2024, 173).

[12] And for Sloterdijk to use repeatedly in *Out of the World* (2024) and *Making the Heavens Speak* (2023) without defining it or acknowledging its genealogy in Schelling and Freud.

the body of his mother is, by implication, "a protozoan monster" and his uterine childhood a fetid, (s)mothering embrace from which he has escaped and to which he does not want to return. Kerans is part of that more general patriarchal process whereby, as Irigaray (1993, 15) puts it, "we have turned the mother into the devouring monster." When he is below he is asked "'How's the grey sweet mother of us all?'" to which he replies "'Feels like home'" (Ballard 1983, 105), but an unhomely or *unheimlich* home which is uncanny.

Kerans descends further into this uncanny home by entering an auditorium "with its blurred walls cloaked with silt [which] rose up above him like a huge velvet-upholstered womb in a surrealist nightmare" (108). Yet Kerans almost dies from diving too deep, but in the process experiences a kind of rebirth and the realisation of his own dream of dissolution of his ego and body boundaries:

> as the spotlight flared across the domed ceiling, illuminating the huge vacant womb for the last time, Kerans felt the warm blood-filled nausea of the chamber flood in upon him. He lay back [...] the soothing pressure of the water penetrating his suit so that the barriers between his own private bloodstream and that of the giant amnion seemed no longer to exist. The deep cradle of silt carried him gently like an immense placenta, infinitely softer than any bed he had ever known. (110)

When Kerans's dream is actualised, it is not as horrific as he feared it might have been, and the embracing, entangling protozoan mother-monster turns out to be an immense, soft placenta with living black waters. Kerans's view of wetlands changes from oral-sadistic, consuming and death-dealing *vagina dentata* to corporeal-pleasurable, (re)producing and life-giving placenta. The silt is no longer horrific and nightmarish, but comforting and nurturing. The velvet-upholstered womb in a surrealist nightmare becomes a deep cradle in a pleasant daydream. The auditorium becomes transformed in recollection into "the great womb-chamber of the planetarium" (112), an exterior architectural analogue, or homologue, of interior bodily space. Kerans is also transformed into what Donna Haraway (1992, 296) calls (and calls herself) "a planetary foetus [...] gestating in the amniotic effluvia of terminal industrialism."

So when Strangman, the villain of the story, undertakes the project of pumping dry a section of London, "turning the once limpid beauty of the underwater city into a drained and festering sewer," Kerans is "unable to accept the logic of the rebirth before him" (Ballard 1983, 121), a rebirth unlike the one which he has just experienced in that instead of returning to the deep, to the amniotic, as he did when he dove into it, part of the lagoon is pumped dry in order to try to return London to its pre-drowned state. For Beatrice, the drained section of London is "'like some imaginary city of Hell'" (123), a necropolis, in fact (151). This Hell is Dantesque, or at least like the fifth circle of Dante's *Inferno* set in the slime of the Stygian marsh, especially with "the black slime oozing" out of the pores of the protozoan mother-monster:

> everything was covered with a fine coating of silt, smothering whatever grace and character had once distinguished the streets, so that the entire city seemed to Kerans to have been resurrected from its own sewers [...] The once translucent threshold of the womb had vanished, its place taken by the gateway to a sewer. (126–127)

As it is in Paris where the marshes have vanished and been replaced by its sewers with their gateways as in Hugo's *Les Miserables* (see Giblett 2016, chapter 2). Vagina and anus in Ballard's post-apocalyptic London are united in the cloaca of the city-mother (53 and 61) as they are for Bonaparte in women more generally. Drowned London had reverted from modern city to primeval swamp and now redrained London reverts from primeval swamp to hypermodern slime and sewer covered in silt in the Anthropobscene.

In accordance with the fundamental structuring principle of the novel whereby inner and outer are analogous, if not homologous, the drained lagoon affects Kerans profoundly and "he began to sink rapidly into a state of dulled inertia, from which he tried helplessly to rouse himself. Dimly he realised that the lagoon had represented a complex of neuronic needs that were impossible to satisfy by any other means" (129). Kerans experiences a sort of post-natal depression, not as mother, but as someone who has just been (re-)born. As a result, he becomes a kind of wetlands conservationist, though hardly an ecofeminist, who is sensitive to the homology between the inner and outer landscapes and aware of the connections between the depths and the surfaces, the repressed and the expressed, the past and the present. Although he is now fascinated with and not horrified by the wetland as womb, he is still not interested in his mother, and the Body of the Mother. The mother (in two senses) is absent from his story, though it/she is virtually present, or has left traces, in the references to the protozoan monster and black water of the lagoons. Despite his transformation from imperialist cartographer to wetlands conservationist, Kerans is still a hero from whose life, whose story, his mother is largely absent. Kerans extends his wetlands conservationism to becoming an eco-saboteur who blows up the dykes which hold the water back and heads south "a second Adam searching for the forgotten paradises of the reborn sun" (175).

Ballard's work is important as an attempt to tell the story of the imperial city as colonising wetlands at 'home and abroad' and to think the unthinkable future albeit as repetition of the past in the return of those repressed wetlands. For Frederic Jameson (1984, 245) "today the past is dead, transformed into a packet of wellworn and thumbed glossy images" in the nostalgia industry. As for the future, he goes on to suggest, "it is for us either irrelevant or unthinkable." Yet for Ballard in *The Drowned World* the future *is* thinkable, but only as a return to the past. This return is not nostalgic in the sense of homesickness, but *unheimlich* or unhomely which is both horrifying and fascinating. What Jameson (1984, 245) calls the "Wagnerian and Spenglerian world dissolutions of J. G. Ballard" should stand as what he goes on to describe as "exemplary illustrations of the ways in which the imagination of a dying class—in this case the cancelled future of a vanished colonial and imperial destiny—seeks to intoxicate itself with images of death." Or perhaps more precisely in the case of *The Drowned World* with images of parodies of life, of new life being born from the wetland as womb.

These images include the destruction, and rebirth, of the world by water in *The Drowned World*. But not just any water, or even water per se, and certainly not the individualist death by water of Eliot's world-weary, modernist and mythopoeically nostalgic wasteland. It is the death of western civilisation by wetlands, by the return of its endo- and exo-colonial repressed, and a death which is a parodic repetition of its birth from the uncanny. The future is thought, not as difference, but as parody of sameness, the future as parodic repetition of the past, both of the species and the individual. The future is thinkable, but only as repetition of or return to the past. There are two kinds of countervailing trajectories at work and in tension here: one entailing a return to the repressed via the uncanny uterus and the other involving the future as a repetition of the uterine past. Perhaps the richness and suggestiveness of Ballard's novel can be attributed in part to the way in which the two lines are used to figure each other.

Part of its fascination can perhaps also be attributed to its quantum interplay with time and space, and with the way they are used to figure each other with colonised places figured as palaeozoic past; post-colonial here and now as pre-colonised and palaeozoic then and there. Ballard's work is for Jameson (1984, 245) "so rich and corrupt" because it:

> testifies powerfully to the contradictions of a properly representational attempt to grasp the future directly. I would argue, however, that the most characteristic science fiction does not simply attempt to imagine the 'real' future of our social system. Rather, its multiple mock futures serve the quite different function of transforming our own present into the determinate past of something yet to come.

In particular, *The Drowned World* transforms our own ostensibly post-colonial and post-modern present into the determinate, and repressed, past of the pre-colonial and uterine yet to come. It enacts the postmodern future anterior, as Lyotard (1984, 79 and 81) puts it, in which the present is transformed back into the determinate past of the pre-modern wetland wilderness yet to come as postmodern wilderness.

10.3 The Lagoon

By contrast with Conrad's story 'The Lagoon,' and by a curious literary historical irony, in Janet Frame's story of the same title as Conrad's the lagoon represents a way for the first person narrator of establishing and celebrating matrilineal descent, particularly a connection back to one of her grandmothers and great-grandmothers. The lagoon is not a particularly pleasant place, it is not romanticised, but nor is it psychogeopathologised, as it is for Conrad, as a place of darkness and horror alone:

> at low tide the water is sucked back into the harbour and there is no lagoon, only a stretch of dirty grey sand shaded with dark pools of sea water where you may find a baby octopus if you are lucky, or the spotted orange old house of a crab or the drowned wreckage of a child's toy boat. There is a bridge over the lagoon where you may look down into the little pools and see your image tangled up with sea water and rushes and little bits of cloud. And sometimes at night there is an underwater moon, dim and secret. All this my grandmother told me, my Picton grandmother. (Frame 1991, 1)

Although the tide sucks the water from the lagoon back into the harbour, this is not an oral sadistic sucking which enacts a fear of being eaten by (the) (earth) mother, but part of the rhythm of the tides, and of life. At low tide the lagoon is not a place of new life, but of death and destruction—the baby octopus who will not survive without water, the shell from a dead crab, the damaged and lost toy left by the kids who play on the edge, who do not venture in like the narrator, the other kids who are anglicised "Christopher Robins with sand between the toes, sailing toy warships and paddling with bare feet in the pools" (6), the clean and proper children. Ultimately, as it transpires, the lagoon is the place where the narrator's maori princess and murderess great-grandmother drowned her husband (6). Yet none of these aspects of the lagoon is dwelt upon, nor psychogeopathologised, but are a part of the rhythms of life and death. Frame is part of the minority tradition which sees the wetland as a place of both death and life, darkness and light, as living black water.

The lagoon is not a place which is feared because the self might be absorbed. Rather it is a place where self and other mingle. It is not the pool of Narcissus where the still reflective surface of the water gives back the image of the self as other so that the self can fall in love with itself as other. Nor is it the artificialised wet land of landscape painting which enables the bourgeois viewer to reflect on the surface of things by reflecting on the still surface of the water. Rather "the water was brown and shiny" (6), the surface of the lagoon broken up by wind and wave which enables one to see both surface (the self) and depth (the other) above and below without falling in love with the self on the surface, nor fearing absorption by the womb of life, nor repressing the depths below, nor desiring sublimation into the heights above. One is situated comfortably and mid-way between surface, depth and height.

The narrator's grandmother instructs her to "see the lagoon," to see it for what it is, not as a psychogeopathologised projection of fears, though it is a place of fear, even more so than the bush (4), but that was part of its fascination. It was "the dirty lagoon, full of drifting wood and seaweed and crab's claws. It is dirty and sandy and smelly in summer" (2). Despite its unpleasantness in terms of the conventions of European aesthetics, the narrator "liked my grandmother to talk about the lagoon [...] But the lagoon never had a proper story, or if it had a proper story, my grandmother never told me" (2). The lagoon cannot speak for itself, it is always spoken on behalf of for better or worse (usually for worse, but invariably situated at one of two extreme poles), as the wetland is more generally in western culture. Frame refuses this either/or logic and deconstructs its binarism to adopt a both/and logic in which the lagoon is both fascinating and unpleasant, rather than fascinating and horrifying.

10.4 Being Born

In a word, lagoons and swamps are places and spaces of what Hannah Arendt (1996, 51, 132–133, 146–148), the eminent philosopher and the student and lover of Martin Heidegger, called "natality," or being born, in her doctoral thesis.[13] Sloterdijk (2020, 158 n13) comments in a note that natality, "which seemingly names something self-evident, does not belong to the vocabulary of philosophy – a treacherous fact"—one pointed out by Irigaray and other feminists. Sloterdijk states that "we find it first" in a book published in German in 1979. Unbeknown to him, Arendt's doctoral thesis on natality was first published in German in 1929, 50 years before.

Arendt placed natality, or being born, at the centre of her social philosophy of life and love as a counter to Heidegger's morbid, solipsistic philosophy of mortality and despair facing the fact of one's own death. In an interview Sloterdijk calls this "the darker side of Heidegger's philosophy," which he "wholly left out" of his discussions, even when he devoted a chapter to the discussion of "Heidegger's politics" (Noordegraaf-Eelens and Schinkel 2011, 191; Sloterdijk 2018, 127–153). Arendt's philosophy of being born (and born again) counters Heidegger philosophy of being towards death in the Anthropobscene. Death is a part of life, but new life is born in an eternal return and cycle of birth, life, death, birth, etc.

Arendt also placed natality and being born again into the public sphere (*polis*) out of the private sphere of the domestic household (*oikos*) at the centre of her political philosophy (see Giblett 2011, chapter 2). Sloterdijk went on to say in the interview that "I try to interpret the movement into life as a kind of permanent birth" (Noordegraaf-Eelens and Schinkel 2011, 191). Natality and being born into the earthly household is also at the center of psycho-political symbiosis and mutuality with the living earth in the Anthropobscene (the topics of the following chapter).

[13] For Arendt's philosophy of love and her relationship with, and views of, Heidegger, see Kristeva (2001, xv, xvii, 8, 44–46, 48, 66–67, 141, 239–240, especially 243 n27 for Arendt's outrageous and scathing "true story of Heidegger the fox," 246 n75) and Eilenberger (2020, 192, 240, 379 n22, 24, plus photos of Heidegger, the sly fox).

References

Arendt, Hannah. 1996. *Love and Saint Augustine*. Edited by Joanna Vecchiarelli Scott and Judith Chelius Stark. Chicago, IL: University of Chicago Press. First published in German in 1929.

Ballard, J. G. 1983. *The Drowned World*. London: Dent.

Benjamin, Walter. 1973. *Illuminations*. Translated by Harry Zohn. London: Fontana.

Benjamin, Walter. 1999. *Selected Writings: Volume 2, 1927-1934*. Edited by Michael Jennings, Howard Eiland and Gary Smith. Cambridge, MA: Belknap Press of Harvard University Press.

Bonaparte, Marie. 1953. *Female Sexuality*. London: Imago.

Cicero. 1972. *The Nature of the Gods*. Translated by Horace C. P. McGregor. London: Penguin.

Conrad, Joseph. 1977. The Lagoon. In *Tales of Unrest*. Harmondsworth: Penguin. First published 1898.

Earll, Robert. 1991. Shorelines and Estuaries: Turning the Tide. In *Save the Earth*, ed. Jonathon Porritt, 158–164. North Ryde: Angus & Robertson.

Easthope, Antony. 1986. *What A Man's Gotta Do: The Masculine Myth in Popular Culture*. London: Paladin.

Eilenberger, Wolfram. 2020. *Time of the Magicians: Wittgenstein, Benjamin, Cassirer, Heidegger and the Decade that Reinvented Philosophy*. Translated by Shaun Whiteside. New York: Penguin.

Frame, Janet. 1991. The Lagoon. In *The Lagoon and Other Stories*. London: Bloomsbury. First published 1951.

Freud, Sigmund. 1973. *Introductory Lectures on Psychoanalysis*. Pelican Freud Library 1. Harmondsworth: Penguin. First published in German in 1917.

Freud, Sigmund. 1984. *On Metapsychology: The Theory of Psychoanalysis*. Pelican Freud Library 11. Harmondsworth: Penguin.

Giblett, Rod. 2011. *People and Places of Nature and Culture*. Bristol: Intellect Books.

Giblett, Rod. 2016. *Cities and Wetlands: Return of the Repressed in Nature and Culture*. London: Bloomsbury.

Gilbert, Sandra, and Susan Gubar. 1989. *No Man's Land: The Place of the Woman Writer in the Twentieth Century, volume II, Sexchanges*. New Haven, CT: Yale University Press.

Gimbutas, Marija. 1989. *The Language of the Goddess: Unearthing the Hidden Symbols of Western Civilization*. San Francisco, CA: Harper and Row.

Haggard, H. Rider. 1991. *She*. Oxford: Oxford University Press. First published 1887.

Haraway, Donna. 1992. The Promises of Monsters: A Regenerative Politics for Inappropriate/d Others. In *Cultural Studies*, ed. Lawrence Grossberg, Cary Nelson, and Paula A. Treichler, 295–337. New York: Routledge.

Heraclitus. 2003. *Fragments*. Translated by Brooks Haxton. New York: Penguin.

Hiebert, Theodor. 2008. Air, the First Sacred Thing: The Conception of רוּחַ in the Hebrew Scriptures. In *Exploring Ecological Hermeneutics*, ed. Norman Habel and Peter Trudinger, 9–19. Atlanta, GA: Society of Biblical Literature.

Hughes, Robert. 1980. *The Shock of the New: Art and the Century of Change*. London: British Broadcasting Corporation.

Irigaray, Luce. 1977. Women's Exile. *Ideology and Consciousness* 1:62–76.

Irigaray, Luce. 1985a. *Speculum of the Other Woman*. Translated by Catherine Porter. Ithaca, NY: Cornell University Press.

Irigaray, Luce. 1985b. *This Sex Which Is Not One*. Translated by Catherine Porter. Ithaca, NY: Cornell University Press.

Irigaray, Luce. 1991. *Marine Lover of Fredrich Nietzsche*. Translated by Gillian C. Gill. New York: Columbia University Press.

Irigaray, Luce. 1993. Body Against Body: In Relation to the Mother. In *Sexes and Genealogies*, 9–21. Translated by Gillian C. Gill. New York: Columbia University Press.

Jameson, Frederic. 1984. Progress Versus Utopia; or, Can We Imagine the Future. In *Art After Modernism: Rethinking Representation*, ed. Brian Wallis. New York: The New Museum of Contemporary Art.

Jardine, Alice. 1985. *Gynesis: Configurations of Woman and Modernity*. Ithaca, NY: Cornell University Press.

Kirk, G. S., J. E. Raven, and M. Schofield, eds. 1983. *The Presocratic Philosophers: A Critical History with a Selection of Texts*. 2nd ed. Cambridge, UK: Cambridge University Press.

Kristeva, Julia. 2001. *Hannah Arendt*. Translated by Ross Guberman. New York: Columbia University Press. First published in French in 1999.

Lacan, Jacques. 1977. *Ecrits: A Selection*. Translated by Alan Sheridan. London: Tavistock.

Lloyd, G. R., and Nathan Sivin. 2002. *The Way and the Word: Science and Medicine in Early China and Greece*. New Haven, CT: Yale University Press.

Lyotard, Jean-François. 1984. Answering the Question: What is Postmodernism? Translated by Regis Durand. In *The Postmodern Condition: A Report on Knowledge*, 71–82. Translated by Geoff Bennington and Brian Massumi. Manchester: Manchester University Press.

Lyotard, Jean François. 1993. *Libidinal Economy*. Translated by Iain Hamilton Grant. London: Athlone.

McComb, A. J., and P. S. Lake. 1990. *Australian Wetlands*. North Ryde: Angus and Robertson.

Miller, David. 1989. *Dark Eden: The Swamp in Nineteenth-Century American Culture*. Cambridge, UK: Cambridge University Press.

Neumann, Erich. 1955. *The Great Mother: An Analysis of the Archetype*. Translated by Ralph Manheim. London: Routledge and Kegan Paul.

Niering, William A. 1966. *The Life of the Marsh: The North American Wetlands*. New York: McGraw Hill.

Noordegraaf-Eelens, Liesbeth, and Willem Schinkel. 2011. The Space of Global Capitalism and Its Imaginary Imperialism: An Interview with Peter Sloterdijk. In *In Media Res: Peter Sloterdijk's Spherological Poetics of Being*, ed. Willem Schinkel and Liesbeth Noordegraaf-Eelens, 185–194. Amsterdam: Amsterdam University Press.

Paglia, Camille. 1991. *Sexual Personae: Art and Decadence from Nefertiti to Emily Dickinson*. New York: Vintage.

Sjöö, Monica, and Barbara Mor. 1991. *The Great Cosmic Mother: Rediscovering the Religion of the Earth*. 2nd ed. San Francisco, CA: Harper.

Sloterdijk, Peter. 1987. *Critique of Cynical Reason*. Translated by Michael Eldred. Minneapolis, MN: University of Minnesota Press. First published in German in 1983.

Sloterdijk, Peter. 2018. *What Happened in the 20th Century*. Translated by Christopher Turner. Cambridge, UK: Polity Press. First published in German in 2016.

Sloterdijk, Peter. 2020. *Infinite Mobilization: Towards a Critique of Political Kinetics*. Translated by Sandra Berjan. Cambridge, UK: Polity Press. First published in German in 1989.

Sloterdijk, Peter. 2023. *Making the Heavens Speak: Religion as Poetry*. Translated by Robert Hughes. Cambridge, UK: Polity.

Sloterdijk, Peter. 2024. *Out of the World*. Translated by Corey A. Dansereau and Gill Zimmermann. Stanford, CA: Stanford University Press. First published in German in 1993.

Sloterdijk, Peter. 2025. *The Terrible Children of Modernity: An Antigenealogical Experiment*. Translated by Oliver Berghof. New York: Columbia University Press. First published in German in 2014.

Thoreau, Henry D. 1993. *Faith in a Seed: The Dispersion of Seeds and Other Late Natural History Writings*. Edited by Bradley P. Dean. Washington, D.C.: Island Press/Covelo, California: Shearwater Books.

CHAPTER 11

Symbiosis, Livelihood, and Bioregion

Abstract Symbiosis, livelihood, and bioregion are ways of living and being with nature in the Anthropobscene and countering sublimation with desublimation in the Symbiocene, the hoped-for epoch superseding the Anthropobscene. A **bioregion** is a geomorphological and biological region, the watershed, the valley, the plain, the wetland, the aquifer, etc. where or on which humans live and work, and which sustains human and more-than-human life. This relationship is situated on a continuum between the mutually beneficial and 'normal' biological and psychological desire for, and pleasures of, symbiosis and mutuality at one end, and the parasitical, psychotic, and failed drive for mastery at the other. Raymond Williams's concept of **livelihood** implies both our work and our physical surrounds, their environmental supports and effects. **Symbiosis** is both a biological and a psychological term, what could be called bio-symbiosis and psycho-symbiosis. After drawing on Lynn Margulis's work on bio-symbiosis, Margaret Mahler's work on psycho-symbiosis, Jessica Benjamin's work on bonding, Michel Serres's work on parasitism and the natural contract, and Raymond Williams's late ecocultural work on livelihood and socialist ecology to make these cases, this chapter and the book then concludes by advocating for psychopolitical symbiosis and mutuality with the living earth in the Symbiocene.

R. Giblett, *Living with Nature in the Anthropobscene*,
https://doi.org/10.1007/978-3-032-12822-5_11

To just plain live on earth and to revalue the meaning and inhabitation of nature, humans need new ways of thinking and acting, new ways of seeing, saying and doing, new ways of being in the Anthroposbscene. The three concepts of bioregion, livelihood and symbiosis have figured prominently in recent discussions about the relationship between humans and habitat. A **bioregion** is a geomorphological and biological region, the watershed, the valley, the plain, the wetland, the aquifer, etc. where or on which humans live and work, and which sustains our life. Raymond Williams's concept of **livelihood** implies both our work and our physical surrounds, their environmental supports and effects. **Symbiosis** is both a biological and a psychological term, what could be called bio–symbiosis and psycho–symbiosis. Lynn Margulis used the term to convey biological mutuality as distinct from parasitism. Margaret Mahler applied the term to child–mother relationships covering a continuum from the 'normal' to the psychotic and for phases of childhood development from union with the mother to separation and individuation. In this chapter I develop a connection between all three by arguing that a **bioregion** always sustains our **livelihood** in relationship with the living earth. This relationship is situated on a continuum between mutually beneficial and 'normal' biological and psychological **symbiosis** and mutuality at one end, and parasitical and psychotic mastery at the other. I conclude the chapter and the book by advocating for the former and calling for mutuality with the living earth.

11.1 Symbiosis

Biologically humans are a community. Indeed, as Theodore Roszak (1992, 154) argues citing Lynn Margulis, "all organisms are 'metabolically complex communities of a multitude of tightly organized beings.'" For her, Roszak (155) goes on to relate, "from the symbiotic point of view, there are no 'individuals' – except perhaps the bacteria. All beings are 'intrinsically communities'." For nearly 40 years Margulis (1998, 33) has been the most cogent proponent of the notion of symbiosis, "the term coined by the German botanist Anton de Bary in 1873" and taken up by Eugene Warming (1909, 83–95) in the early twentieth century. Margulis (1971, 49; see also 1981, 161) defines symbiosis as "the living together of two or more organisms in close association. To exclude the many kinds of parasitic relationships known in nature, the term is often restricted to associations that are of mutual advantage to the partners."

Symbiosis is also one of the key concepts of bioregionalism (discussed in greater detail later). One of the features of the bioregional paradigm for Kirkpatrick Sale (1985, 50 and 112), particularly as it relates to 'society'[1] and also presumably in relation to the natural environment, is symbiosis, or, as Sale defines it, "biological interaction" and "mutual dependence as the means of survival." Symbiosis in biological terms operates primarily at a microbiological, interspecies level, though it has wider macrobiological pertinence between animals and plants. Sale (1985, 113) extends its range even further and uses it as a model for "a successful human society" which would operate symbiotically at the macrobiological, or bioregional level, particularly when it comes to cities. Yet Sale is using symbiosis in the restricted or narrow sense of mutual benefit as all animal species (including humans) are always already in symbiosis in a broader sense with the natural environment to greater or lesser extent, with greater or lesser mutual dependency and benefit. Human beings are, whether we like it or not, in symbiosis with the oxygen–producing plants of this planet. Every breath of air we breathe re–affirms this symbiosis. As Margulis (1998, 5) puts it, "we are symbionts on a symbiotic planet." Or as Sloterdijk (2020, 191) puts it, "humans are beings who can flourish only in the element of reciprocal support."

Parasitism can be and has been a model for a 'successful' human society which survives, however parasitically, to the detriment of other species, habitats, ecosystems and biomes. Sale wants symbiosis to be a model (even a realisable utopian model) for a human society which is mutually 'successful' or sustainable with its bioregion. Rather than being a model for or a metaphor of a successful or better human society, symbiosis is the current state of play as humans are organisms, are biological creatures. Mutualism may be a useful model for a better human society, though all appeals to nature for a model of society are at best fraught with danger as nature is a cultural construct and at worst imbued with class politics for nature is an object of capitalist exploitation. Appeals to nature are to a court whose only recourse for further appeals is to refer the case on to the higher court of culture where the case is argued (and won or lost) on cultural grounds.

[1] "Society," as Sloterdijk (2024, 480 n112) points out, is "the object of classical sociology [...] as an association of subjects." He also states with sardonic irony that "society" is "that real-life fabulous creature, [...] an actually existing million-headed fantasy creature" (Sloterdijk 2016, 5, 6).

A mutualistic theory of human society is all very well in theory, but when it comes to putting it into practice it is another matter, especially in relation to that most parasitic of beasts or even monsters, the city. It is easy to be long on generalities and short on specifics on how the city might living mutually and not parasitically in symbiosis with its bioregion. It is easy to diagnose the parasitic ills of the city, but much harder to get the city to take its mutualistic medicine. Cities, Sale (1985, 65) argues,

> particularly modern industrial cities, are like colonizers, grand suction systems drawing their life from everywhere in the surrounding nation, indeed the surrounding world [...] The contemporary high–rise city, in short, is an environmental parasite as it extracts its lifeblood from elsewhere and an environmental pathogen as it sends back its wastes.

It is difficult to see how such cities might cease to be parasites on their bioregion and live mutually with it. Smaller scale, decentralised communities may be able to do so.

The modern city is founded and continues to depend on a parasitic relationship with its bioregion. Despite the parasitism of the modern city, it is still in a symbiotic relationship with its bioregion and with the earth. Perhaps, as Sale suggests, it is in a pathologically symbiotic rather than healthily symbiotic relation with the world. In Margaret Mahler's terms, it is in psychotic symbiosis rather than 'normal' symbiosis. For Mahler (1972, 333) "'growing up' entails a gradual growing away from the normal state of human symbiosis, of 'one–ness' with the mother." Similarly, modernity entails a gradual growing away from the 'normal' or traditional state of human symbiosis, of 'one–ness' with the earth. Modernity is thus a pathology, a land pathology and a psychopathology, a psychogeopathology. Indigenous people's oneness with the earth is the norm from which modern people have deviated pathologically, both physiopathologically and psychopathologically.[2]

Mahler (1968, 6) could be described as a neo–Freudian psychoanalyst whose work on what she calls "the symbiosis theory of the development of the human being" spans over three decades beginning in the 1950s. In this theory there are basically three phases of personality development: the normal autistic; the normal symbiotic; and the separation–individuation

[2] I discuss the possibility of the sustainable city (and make some proposals for it) in Giblett (2009, chapter 3).

phases.[3] These phases refer primarily to "the development of object relationship," especially with the mother and corresponding roughly to Freud's oral, anal and phallic stages. Mahler's (1968, 2) theory begins from the foundation of what she calls "that vital and basic need of the human young in his [*sic*] early months of life: symbiosis with a mother or mother substitute." For Mahler et al. (1975, 44; see also Mahler 1968, 8), normal symbiosis is the phase in which "the infant behaves and functions as though he [*sic*] and his mother were an omnipotent system – a dual unity within one common boundary," just as the bioregion normally is the dual unity of indigenous humans with the earth within one common boundary.

The autistic child for Mahler develops a psychotic defense against the lack of this need. The psychotic symbiotic child is thus in the terms of Mahler, Pine and Bergman (1975, 12) "unable to use the mother as a real external object as a basis for developing a stable sense of separateness from, and relatedness to, the world of reality." Similarly the psychotic city–dweller is generally unable to see the earth as separate and is unable to relate to her other than as Mother Nature in the packaged commodities of nature documentaries, national parks, processed food, tourist destinations, and so on.

For Mahler and Furer (1966, 559) this lack of separation in "symbiotic child psychosis" involves "a regression to the stage of object relationship that exists before self and other representation have been distinguished." In other words, symbiotic psychosis entails a regression to the symbiotic state of early infancy, even to the parasitic, or perhaps more precisely, intrauterine state, in which the infant is unable to distinguish between self and (m)other. Following Helene Deutsch, Mahler (1952, 286) argues that "the intrauterine, parasite–host relationship within the mother organism must be replaced in the postnatal period by the infant's being enveloped, as it were, in the extrauterine matrix of the mother's nursing care, a kind of *social symbiosis*." Mahler (1968, 34) repeated the same view 16 years later. Yet it is arguable whether the intrauterine relationship is strictly parasitic as the foetus does not gain considerably at the mother's expense. It is perhaps more precisely inquilistic (see below) as the uterus shares the body

[3] Sloterdijk (2024, 4, 29) argues that "the phase theories of developmental psychology [including Mahler's] usually begin with the individual in the state of separation from the mother." He concludes later in the same paragraph that "the dissolution of the postnatal symbiosis with the mother makes it possible to speak of an effective individuation." Mahler's theorisation of the phases of developmental psychology begins with the postnatal symbiosis with the mother.

of the mother usually without significant disadvantage to the mother in the short-term, but often in the long run with a prolapse and the need for surgical intervention, including a hysterectomy.

Just as the symbiotic psychotic infant is unable to distinguish between self and (m)other, so the symbiotic psychotic city–dweller is unable to distinguish between self and the earth. This regression is thus, Mahler and Furer (1966, 560) argue, "an archaic defence mechanism, a restitutive attempt that, by way of the delusion of oneness with the 'mothering principle', serves the function of survival." This delusion of oneness entails delusional feelings of omnipotence so for Mahler within this sense of oneness lie the seeds of its own destruction. Later she goes on to argue with Pine and Bergman (Mahler, Pine and Bergmman 1975, 45; see also Mahler 1968, 9) that "the essential feature of symbiosis is hallucinatory or delusional somatopsychic *omnipotent* fusion with the representation of the mother and, in particular, the delusion of a common boundary between two physically separate individuals."[4] The psychotic city dweller has a delusional oneness with the mothering principle of nature, and a delusional omnipotence over it/her, through the produced and packaged commodities of the hypermarket shelf and the electronic media which serve the function of survival (and to that extent is 'successful'), but not the function of living symbiotically, sustainably and spiritually with the earth.

The symbiotic psychotic, Mahler and Furer (1966, 560) go on to argue, does not cling to the mother *per se*, but to "a hypercathected, yet at the same time devitalized and deanimated concrete symbol which he [*sic*] substitutes for her – a psychotic transitional object to which he constantly resorts in stereotyped fashion," what she also calls "a 'psychotic fetish'." The living body of the earth, nurturing and horrifying, life–giving and death–dealing becomes the dead matter of Mother Nature, an industrial resource to be extracted and exploited, a commodity to be bought and sold, produced and consumed, a bauble to be fetishised by modern communication and information technologies in advertising, documentaries and tourism and a waste dump to absorb and hide garbage. As Sloterdijk (2018, 9) says, "nature-as-background is nowadays construed as an integral storehouse of resources and as a universal landfill."

The modalities of this relationship between subject and object give rise to what Mahler (1952, 292) calls "the brittle ego of the 'symbiotic psychotic child'" who experiences the world as "hostile and threatening

[4] The word 'physically' is added to the later version.

because it has to be met as a separate being." The response of the symbiotic psychotic, the fragile, hard–edged ego of the city dweller, to this world is to phantasise it precisely as **not** a separate being. So much for the symbiotic psychotic. What of the normal symbiotic?

The normal symbiotic child in Mahler's sense would presumably be the isolated individualist who is constituted as subject insofar as the earth is constituted as object. These terms need to be reconstrued critically in psychopolitically ecological terms. The 'normal' symbiotic city dweller (if such a thing is possible) or the normal symbiotic bioregion dweller, would use the bioregion as a real external subject in her own right, as Great Goddess, for developing a stable sense of inter–subjective interaction and mutual aid with her. The normal symbiotic has developed from and cannot return to the symbiosis of early infancy.

In theorising the normal symbiotic phase of early infancy Mahler (1968, 7) is quite explicit that "the term symbiosis is borrowed from biology, where it is used to refer to a close functional association of two organisms to their mutual advantage." Yet if there is mutual advantage in the symbiosis of early infancy it is not equal as the infant requires the mother or mother substitute to survive biologically whereas the mother does not require the infant for her to survive biologically. In biological terms the relationship is strictly commensal—benefit to one, no harm to the other. Infantile and childhood development can be seen to go through two distinct symbiotic phases: first, an intrauterine period of inquilinism where, in the words of a dictionary of biology, "one party shares the nest or home of the other without significant disadvantage to the 'owner'" in contrast to parasitism "where one party gains considerably at the other's expense"; secondly, a postnatal period of commensalism "where one party gains some benefit [...] while the other suffers no serious disadvantage" (Abercrombie et al. 1990, 542–543). A third stage of mutuality where both parties benefit and neither suffers would perhaps ideally be achieved and maintained in later childhood and adulthood with the proviso mentioned earlier of a possible prolapse and/or hysterectomy.

Yet rather than using the term symbiosis in a strict biological sense, Mahler Pine and Bergman (1975, 44; their emphases; see also Mahler 1968, 9) go on to argue that:

> the term *symbiosis* in this context is a metaphor. Unlike the biological concept of symbiosis, it does not describe what actually happens in a mutually beneficial relationship between two *separate* individuals of different species. It describes that state of undifferentiation, of fusion with mother, in which the 'I' is not yet differentiated from the 'not–I' and which inside and outside are only gradually coming to be sensed as different.

Yet symbiosis in the normal symbiotic relationship could describe what actually happens in a mutually beneficial relationship between two separate individuals, or more precisely communities, of different species living in a bioregion. Symbiosis would no longer be a mere 'metaphor,' or a heuristic device or explanatory trope for something else, but a lived relationship. Rather than being in a state of fusion with the earth, humans in bioregional symbiosis would experience themselves as members of a community interacting in mutual benefit with other species within a common bioregion, indeed, with the whole earth, and with the whole earth considered as a living organism.

Mahler's work has been critiqued for this dubious premise of an undifferentiated, symbiotic union between mother and child (see Klein 1980). In the 1980s Daniel Stern (1985, 10) contended bluntly that "infants never experience a period of total self/other undifferentiation." Analogously the indigene is never totally undifferentiated from or completely symbiotic with the earth. In the loving union of people and earth, Aboriginal people are not subsumed beneath, or swallowed up by, or completely identified with, the earth to the point of losing their own separate identity. People and earth are subjects in their own right, just as the mother is a subject in her own right. Jessica Benjamin (1990, 20; see also Plumwood 1993, 154–160) picked up and developed Stern's position and argued that

> the other whom the self meets is also a self, a subject in his or her own right [...] that other subject is different and yet alike [...] the idea of intersubjectivity reorients the conception of the psychic world from a subject's relations to its object toward a subject meeting another subject.

The earth whom the self meets is also a self, a subject in its or her own right.

Although the earth is a subject like the self, the earth is not the same as the self. The earth is different and yet alike. Jessica Benjamin (1990, 47) argues that

> sameness and difference exist simultaneously in mutual recognition [...] self and other are not merged [...] the externality of the other makes one feel one is truly being 'fed,' getting nourishment from the outside, rather than supplying everything for oneself.

In mutual recognition the earth is different from and the same as people. Aboriginal people and the earth are not merged into an oceanic feeling of complete undifferentiation or loss of mutually distinct identity (see Giblett 2011, chapter 12). The externality and recognition of the earth makes one feel one is truly being fed by nourishing terrains rather than fed by consuming land, by eating earth.

The idea of intersubjectivity reorients the conception of the psychic world away from a subject's relations to its object toward a subject meeting another subject. The earth is another subject as Jessica Benjamin (1990, 23) puts it in relation to the mother:

> The idea of mutual recognition is crucial to the intersubjective view [...] the mother is a subject in her own right [just as the earth is, or should be, a subject in her own right] [...] She is external reality [the earth is external reality] – but she is rarely regarded as another subject.

The earth is rarely regarded as another subject despite its generosity, despite our lack of gratitude towards it and despite the lack of reciprocity in our relationship with it/her. Yet as Serres (1982, 8) puts it "the parasitic relation is intersubjective." The parasitic city and the parasitic modern city-zen constitutes itself as subject and its host, the earth, as subject. The psychotic city dweller, by contrast, has a hypercathected and delusional sense of oneness with the earth that denies it/her subjectivity. Despite this difference, the flow of goods between the earth and the psychotic and parasitic city dweller is one way as is the flow of bads back.

A similar imbalance pertains in the relationship between the child and its parents. Jessica Benjamin (1990, 14) argues that "mutuality [...] persists in spite of the tremendous inequality of the parent–child relationship." Mutuality persists in spite of the tremendous inequality of the earth–indigene, and the even greater inequality of the earth–modern city–dweller relationship. Modern mastery has tried to supersede traditional mutuality and to reverse the tremendous inequality of the earth–indigene relationship. Despite the inequality of the modern–earth relationship, the former is still in symbiosis with the latter.

All people are parasites on the earth to a greater or lesser extent. Serres (1982, 24) argues that:

> man [*sic*] is the universal parasite, everything and everyone around him is a hospitable space. Plants and animals are always his hosts; man is always necessarily their guest. Always taking, never giving. He bends the logic of exchange and of giving in his favour when he is dealing with nature as a whole. When he is dealing with his kind, he continues to do so; he wants to be the parasite of man as well.

Human beings are always the guest of the earth, other animals and plants the host, but there is a world of difference between the oral and anal sadism of mining and pastoralism and the mutual recognition and symbiosis of Aboriginal people and the earth, between the mediated relationship and interaction of the former, and the immediacy of the other.

There is also a difference between indigenes and what Serres (1995, 120 and 122) calls astronauts who live "off–ground" in the extraterrestrial world of cities: "we have all become astronauts, completely deterritorialized [...] Astronaut humanity is floating in space like a foetus in amniotic fluid, tied to the placenta of Mother Earth by all the nutritive passages." Astronaut humanity is floating in the womb of the biosphere tied to the earth mothership at the mouth and the anus and bathing in the amniotic fluid of the atmosphere and the electomagnetosphere inside the skin of orbital extraterrestrial space. Astronaut humanity is not standing with two feet planted firmly on, and rooted in, the ground, drawing nourishment from it and its goods and giving back gratitude and respect. Astronaut humanity ingests good things from the earthly household, digests them and then excretes bad things back into it. According to Milton Klein (1980, 91), Mahler's concept of symbiosis is based on "the image of an unsevered umbilical cord." All humanity has "an unsevered umbilical cord" attaching it symbiotically to the earth, but astronaut humanity is attached parasitically to the earth sucking out nutrients and giving back wastes, and not giving back gratitude and care.

Astronaut humanity is a parasite who, as Lawrence Schehr puts it in his 'Translator's Introduction' to Serres (1982, x), "takes without giving and weakens without killing. The parasite is also a guest, who exchanges his talk, praise and flattery for food." The astronaut for Sloterdijk (2008, 97) is "the Earth's vicar." A vicar is minor clergy and a low-level religious functionary parasitic on the congregation of assembled souls who communes between them and a higher authority with physical and spiritual solid and liquid sustenance. Astronaut humanity does the same with other humans and with more-than-human beings with the commodified and processed goods of nature under the higher authority of industrial capitalists. "The basic figure of modernity" for Sloterdijk (2011, 576) "is the replacement of the all-effectuating God by all-sweeping capital." The fundamental event of modernity is the drive for the conquest of the earthly household as commodity, which has only led to its ruination.

The city and the citi–zen is a parasite which takes from the earth without giving back anything except rubbish and weakens the earth without killing it—yet—for that would be to kill itself. The parasite cannot live without the host; the host can live without the parasite. The city and the citi–zen at home or abroad is also a guest who exchanges his praise and flattery of the earth for sustenance. Perhaps more precisely, a parasite is what Serres (1982, 8) calls "an abusive guest." For Serres (1982, 168), "everything begins with what I call abuse value. The first economic relation is of abuse." Relations of use and use value are pre–economic, or certainly pre–capitalist. The agricultural, manufacturing, mining, and pastoralist industries are instances of industrial capitalist land **ab**use. Serres (1995, 36) goes on to argue later that "the parasite routinely confuses use and abuse. The parasite would destroy the host without realizing it." And him or herself in the process.

Without mutual recognition humans cease to exist and the earth ceases to exist too. Jessica Benjamin (1990, 39) argues that

> if we fully negate the other, that is, if we assume complete control over him [or her] and destroy his [or her] identity and will, then we have negated ourselves as well. For then there is no one there to recognize us, no one there for us to desire.

If humans fully negate the earth, that is, if humans assume complete control over her and destroy her identity and will, then humans have negated ourselves as well. For then there is no one there to recognise us, no one there for us to desire. As Serres (1995, 34) puts it, "former parasites have to become symbionts [...] This is history's bifurcation: either death or symbiosis." The choice is simple: either no future of death or a future of symbiosis. Margulis (1998, 111) concurs that "without 'the other' we do not survive."

Mutual recognition is not a moral stricture, something that humans should do or have to do out of a sense of moral obligation. Rather it is a fact of life, or perhaps more precisely a fact for life as without it humans are dead meat. Jessica Benjamin (1990, 40) insists that "mutual recognition cannot be achieved through obedience, through identification with the other's power, or through repression. It requires, finally, contact with the other." Mutual recognition between people and earth cannot be achieved through obedience to environmental laws or to the figure of the Great Goddess, nor through identification with the Great Mother's power, nor through repression of desire.

Nor even through a natural contract along the lines proposed by Serres (1995, 38) that "we must add to the exclusively social contract a natural contract of symbiosis and reciprocity." Mutual recognition between people and earth requires contact (rather than contract) with the other, with the earth in mutual reciprocity and symbiosis. Not only seeing the earth, but also tasting (and not just eating) the earth, listening, smelling and touching it—or her. To the social and natural contracts the natural contact of symbiosis and reciprocity needs to be added.

That contact can not only be sensual but also erotic. Rather than an undifferentiated relationship with the mother, Jessica Benjamin (1990, 29) suggests that the best model for mutual recognition is erotic union:

> in erotic union we can experience that form of mutual recognition in which both partners lose themselves in each other without loss of self; they lose self-consciousness without loss of awareness. Thus early experiences of mutual recognition already prefigure the dynamics of erotic life.

The loving oneness of Aboriginal people and earth is an erotic union in which both partners lose themselves in each other without loss of self; losing self-consciousness without loss of awareness. This loving oneness and the dynamics of erotic life may prefigure mutual recognition between modern people and earth. Dreaming ecology is erotic ecology. The bioregional dweller is a nature lover who knows intimately his or her home habitat in what could be called an erotic ecology involving intimacy, pleasurable play, sensual appreciation through all the senses and their creative expression in many media (see Giblett 2009, chapter 9). S/he is also what

Felix Guattari (1989, 2000) calls a third ecologist who gains their livelihood sustainably from, and with, the living earth.

11.2 Livelihood

Although the contribution of Williams to the formation of cultural studies is acknowledged in the standard histories, even to the point of being seen as one of its 'founding fathers', the ecological and environmental aspects of his work are ignored (see, for example, Turner 1990), whereas he should be acknowledged as one of the 'founding fathers' of ecocriticism as well. Williams's (1973) *The Country and the City* is the foundational text of ecocriticism in general and political ecocriticism in particular. Cultural studies, with the notable exception of Williams, has suffered just as much from the 'two cultures' divide between the humanities and the sciences and from the culture/nature binary and hierarchy as the sciences. The green Williams is one way of crossing the great divide, deconstructing the binary and decolonising the hierarchies of culture over nature, of science as the master discourse of nature, and of the discourse of technology over the humanities.

In his study of Williams's life and work Fred Inglis (1995, 14) only mentions the green Williams once and largely dismisses this aspect of his work in his 'Prologue' when he condescendingly quotes Williams's eldest daughter in passing as claiming him, "not quite unexpectedly, for green politics, for ecology; for Gaia." In terms of Williams's work over at least the last sixteen years of his life, it was entirely apposite and appropriate for her to do so as we will see below. The intellectual and political sons (including myself) and the biological daughter struggle over the legacy of the father to claim it for ecology, or socialism, or cultural studies, or communication studies, or all four. That his work is engaged in all these fields is its major strength and indicative of his enormous and still pertinent contribution (to all four[5]).

Developing his earlier work on the country and the city and acknowledging that the distinction he drew between them was too stark and that they are, in fact, "indissolubly linked," Williams (1984, 209, 1989b, 227) developed and used the concept of livelihood as a term that mediates both. Livelihood takes place in both the country and the city. The livelihood of the inhabitants of both are dependent and impact on the

[5] I discuss Williams' work on communications and its technologies in Giblett (2008).

livelihoods of inhabitants of the other. Livelihood also deconstructs the culture/nature binary and hierarchy of culture over nature. It is both cultural and natural. Livelihood decolonises the oppression of the natural environment. Livelihood implies both one's work and one's physical surrounds, their environmental supports and effects, as well as something like the American concept of a bioregion, one's geomorphological and biological region, the watershed, the valley, the plain, the wetland, the aquifer, etc. where or on which one lives and works, and which sustains one's life, the lives of indigenes before one and still does so now in places (not to forget the resource regions exploited elsewhere) and the life of other species of fauna as well as flora, and on which one impacts environmentally.

Ecology for Williams was not the ecology of a small minority of nature lovers or accredited experts. These are first and second ecology that Felix Guattari critiqued. Williams extended ecology into what Guattari (1989, 2000) called a third, generalised ecology that "questions the whole of subjectivity and capitalist power formations," an ecology that I am calling psychopolitical ecology following Thoreau and Walter Benjamin (Giblett 2025) and developing what I called postmodern, political ecology (Giblett 2011, chapter 2). For Williams (1985, 214–215; see also 1989b, 210–226):

> what is now known as the ecological argument should not be reduced to its important minor forms; the dangerously rising scale of industrial and chemical pollution; the destruction of some natural habitats and species. The case of the argument is very much harder [...] What is really at issue is a version of the earth and its life forms as extractable and consumable wealth. What is seen is not the sources and resources of many forms of life but everything, including people, as available raw material, to be appropriated and transformed. Against this, the ecological argument has shown, in case after case, and then as a different way of seeing the whole, that a complex physical world and its intricate and interacting biological process cannot for long be treated in such ways, without grave and unforeseen kinds of damage.

For Williams the only way to mount a revolution against the capitalist colonisation and drive for mastery of nature is to produce "a new social and natural order" via the concept and practice of "livelihood." Williams (1984, 219, 1989b, 237; see also 1985, 266–267) warns that:

> the deepest problems we have now to understand and resolve are in [the] real relations of nature and livelihood [... T]he central change we have to make is in the received and dominant concept of the earth and its life forms as raw material for generalized production [...] In order to do so] it is important to avoid a crude contrast between 'nature' and 'production,' and to seek the practical terms of the idea which would supersede both: the idea of 'livelihood,' within, and yet active within, a better understood physical world and all truly necessary physical processes. Both industrial and agrarian capitalism have overridden this idea of livelihood, putting generalized production and profit above it.

The country is not the last bastion of nature against exploitation by capitalism, nor the final refuge of nature in flight from capitalism, but its happy hunting (and gathering, and farming) ground in agricultural and industrial capitalism. Industrial capitalism was and still is just as active in and with the country as it was and is in and with the city.

Livelihood for Williams is a concept and practice which cuts across the rural/country (though these cannot be simply equated as Williams argues), the capital/city, and nature/culture distinctions and divisions. Livelihood deconstructs the culture/nature binary by showing how the relationship between them is hierarchical with the former privileged over the latter. Culture is valorised over nature in culturalism and nature is denied and repressed; nature is exploited and oppressed in and by capitalism and nation–states. The valorisation of nature in naturalism is merely the flip side and reactionary mirror image of culturalism. For the distinction to be made between culture and nature and the relationship between them to be established a third party and term has to be excluded. The excluded third between culture and nature is livelihood. Livelihood is both cultural and natural. There is no livelihood that is not both cultural and natural. Livelihood is cultural and natural.

Livelihood decolonises the distinction between culture and nature, and the unsustainable exploitation of the country by the city and of nature by capitalism. Capitalism is intimately tied up with 'capital–ism,' the fixation on and privileging of the capital city, whether it be of the nation, state or province, over the margins, outskirts and outlying regions. Livelihood empowers the country to resist its exploitation and oppression by the city through creating a sense of place and a viable local economy based on local produce and local markets.

Furthermore, livelihood unites city and country in common dependence and impact on a bioregion. Livelihood takes place in a particular bioregion. Country and city are part of a bioregion. Neither country nor city can survive without their bioregion. Both are dependent, if not parasitic, on it. Agrarian capitalism is parasitic on the country, and industrial capitalism based in the city is parasitic on the country. Agriculture for Michel Serres (1982, 167) is that "old primary parasitism [that] is eliminated by parasites of a superior level, accustomed to noise, the parasites of the megalopolis." The city parasites the country and the country parasites the bioregion. The city is doubly parasitic on its bioregion.

11.3 Bioregion

The concept of bioregion was developed in the 1970's. In less Greco–Latinate terms and more plain Anglo–Saxon one's Sale (1985, 43; see also 2001) suggests that bioregion literally means "life–territory." Even plainer and simpler is 'home land' rather than 'homeland'. The disjunction between the two words signifies uncoupling the conjunction between nationalistic attachment to, and sentiment evoked by, the father– or mother–land. Elaborating on these basic definitions Thomas Berry (1988, 166; see also 86) has defined a bioregion as "an identifiable geographical area of interacting life systems that is relatively self–sustaining in the ever–renewing processes of nature." The bioregion is simply home-habitat.

Extending the concept much further to consider human habitation become problematic. Sale (1985, 55) goes on to define a bioregion in more detail as:

> any part of the earth's surface whose rough boundaries are determined by natural characteristics rather than human dictates, distinguishable from other areas by particular attributes of flora, fauna, water, climate, soils, and landforms, and by the human settlements and cultures those attributes have given rise to.

Sale's definition of a bioregion raises a number of questions such as: which humans and whose dictates would determine the rough boundaries of a bioregion? The fact that Sale can distinguish between boundaries determined by "human dictates" and by "natural characteristics" presupposes humans already distanced from their bioregion and perhaps implies modern humans and their dictates. Permeable boundaries between Australian Aboriginal language and cultural groups are by no accident bioregional boundaries. Sale implicitly includes the role and work of indigenous peoples who have not only been given rise to by the bioregion (which smacks of biological determinism), but have actually shaped and produced the bioregion.

Without explicitly including the perceptions of and uses by indigenous inhabitants, the concept of bioregion can smack of a new doctrine of *terra nullius* with new settlers as new explorers and colonisers. It can be seen as some sort of attempt to get back to a bedrock of biological reality outside indigenous work of the land that writes out the impact of indigenous inhabitants in shaping and producing a bioregion. Williams's concept of livelihood is more inclusive in this respect as it deconstructs the capitalist culture/nature binary opposition and decolonises the European human being/environment, city/country dualisms. It situates humans within the context and the supports which sustain their (our) life.

Bioregion is associated with a "bioregionalism" movement whose threefold aims are, as Jim Dodge (1990, 10) lists them: "a decentralized, self-determined mode of social organization; a culture predicated upon biological integrities and acting in respectful accord [with them]; and a society which honors and abets the spiritual development of its members." Bioregionalism as a political philosophy has been critiqued on the basis that it is only a viable economic option for those "new settlers" who have the financial and other resources to leave the city and relocate in a smaller rural community. Bioregionalism is also part of a much larger social phenomenon whose practitioners may not conceptualise or articulate what they are doing in those terms. Urban baby boomer professionals undergoing a midlife crisis and seeking a seachange brought about by their disenchantment and disillusionment with the alienating city and with its and their 'lifestyle' ('rat race') escape to the country on a journey of self-discovery to find authenticity, community and themselves.

To this extent bioregionalism is nostalgic and utopian (utopia as nostalgia), even reactionary as Andrew Ross (1994, 11) has warned:

> in enshrining the principles of small-scale economies, decentralization, and face-to-face democracy of eco-anarchist thinking, bioregionalism looks back to a low density, rural-tribal past when communities 'coincided' geographically with bioregions defined by watersheds, or microclimates, or biomes […] Above all, bioregionalism embodies, as a political and cultural policy, fundamentalist strains of biological determinism that run unevenly throughout mainstream environmentalist thought.

In its ideas of bioregions giving rise to cultures and settlements in which the latter are merely products of the former (autochthonous) rather than being in a dialectic, or better dialogic, or even better mutually beneficial, relationship with its bioregion bioregionalism is biologically deterministic. Nevertheless, all urban dwellers live in a bioregion anyway irrespective of whether or not they subscribe to bioregionalism as a political philosophy and respond to their midlife crisis by leaving the city to find solace in the country; even Andrew Ross lives in a bioregion; even New York is in a bioregion (see Sanderson 2009).

The shift from bioregion to bioregionalism, from an acknowledgement of the bio– and geo–features of one's region to a fully–fledged political philosophy is fraught with danger. Yet the baby of the bioregion does not need to be thrown out with the bathwater of bioregionalism. The value of bioregion lies not only in making connections with a larger sense of place in which one lives and on which one depends, but also in reconstruing a sense of community from a narrow, stultifying human community to a broader and richer sense of community of all beings, what I am calling a bioregional home-habitat. For Thomas Berry (1988, 166), "the bioregion is the domestic setting of the community." The bioregion is the communal home, home land.

The bioregion has also been defined by Charlene Spretnak (1997, 104):

> the land masses of Earth are organized into bioregions delineated by watersheds (drainage areas) of the river systems or other natural demarcations. Everyone lives in a bioregion and in the Earth's commons [...] The health of communities and nations is dependent on the health of the bioregions.

The boundaries of the bioregion are mapped on the earth's surface by the curving lines of water flows. They are not mapped in the straight lines of national and state borders, but cut across them. No one can escape their bioregion and the commons of land, air and water in which everyone lives and on which everyone depends. The bioregion is the community of living beings and non–living things.

"The community" for Sale (1985, 62) is "the single basic building block of the ecological world." He defines it as "a self–sufficient and self–perpetuating collection of different species that have adapted as a whole to the conditions of their habitat." Indigenous inhabitants of a bioregion living in a hunting-gathering culture would exemplify this sort of community. For indigenous communities the bioregion is their home. The bioregion is the domestic setting of the indigenous community. Non-indigenous residents, especially city-dwellers, are usually locked into a community/individual divide. For them the home is the domestic setting of the individual, the bioregion is the background, the noise, the resource, the setting, not the home of the community. Bioregionalism seeks to change the order of priorities between city and country, to see, and live in, the bioregion as home. The concept of bioregion could extend the boundaries of home beyond the limits of one's piece of land to the edges of the geomorphic and vegetative region in which one lives and in which indigenous inhabitants lived before one (the settler).

The concept of bioregion could even extend the boundaries of home not only outwards from the house and the garden, but also inwards to include humans and the body. As David Suzuki (1998, 59) points out:

> boundaries cannot be drawn between us and air, water, or soil – in fact, we are, quite literally, air, water, soil, and sunlight [...] Whatever we do to the air, water, soil, energy, and biodiversity, we also do to ourselves because there is no separation [...] the way we act towards the environment is the way we act towards ourselves.

A new environmental golden rule: do unto the environment as you would be done to, for you are your environment. After all, humans are predominantly water and humans do grow intestinal flora. In Suzuki's later book with Amanda McConnell (1997) the ways in which humans are our environment are discussed more extensively.

Bioregion extends the boundaries of home outwards and inwards. Yet they are not usually extended far enough outwards to include all that humans make home, or call home, or need for home. Human settlements and cultures have increasingly extended their boundaries beyond the earth's surface to take in more of the earth's depths in mining and more of the earth's heights beyond the mountains and the atmosphere into the electromagnetosphere and orbital extraterrestrial space by way of communication technologies such as radio, television and satellite telecommunications. After the surface of the earth was colonised, the depths and heights are being increasingly colonised. These regions have not yet become regarded or cared for as home, though cyborgs live in them and depend on them. These regions are not only part of the earth–household, but also part of the ecoregion. Ecoregion is the more apposite term today to include not only the bioregion on the earth's surface and heights in the atmosphere in which all beings live and on which they depend, but also the regions of the earth's depths and heights, the ---------ospheres, in which cyborgs and symbionts live and on which they depend, including other bioregions from which they 'source' goods in trade (see Haraway 1995, xi–xx; Giblett 2008, chapter 9).

11.4 Living Earth

Symbiosis is a fact of evolution: no evolution without symbiosis. Symbiosis is a factor in evolution: species evolve through symbiosis with other species. It is also a factor in bioregions, and bioregionalism. Darwin's theory of evolution by natural selection was not possible without bioregions and symbiosis. This theory was a Copernican Revolution in biology that ranks alongside the Marxian revolution in history and political economy and the Freudian revolution of psychoanalysis in psychology. All three decentred human subjectivity.

Louis Althusser (1984, 170–171) acknowledged the pivotal role of Freud and Marx, whilst overlooking Darwin, though Freud himself did not, but overlooked Marx, in his *Introductory Lectures on Psychoanalysis*:

> Humanity has in the course of time had to endure from the hands of science two great outrages upon its naive self-love. The first was when it realized that our earth was not the center of the universe, but only a tiny speck in a world-system of a magnitude hardly conceivable; this is associated in our minds with the name of Copernicus, although Alexandrian doctrines taught something very similar. The second was when biological research robbed man of his peculiar privilege of having been specially created, and relegated him to a descent from the animal world, implying an ineradicable animal nature in him: this transvaluation has been accomplished in our own time upon the instigation of Charles Darwin, Wallace, and their predecessors, and not without the most violent opposition from their contemporaries. But man's craving for grandiosity is now suffering the third and most bitter blow from present-day psychological research which is endeavoring to prove to the ego of each one of us that he is not even master in his own house, but that he must remain content with the veriest scraps of information about what is going on unconsciously in his own mind. We psycho-analysts were neither the first nor the only ones to propose to mankind that they should look inward; but it appears to be our lot to advocate it most insistently and to support it by empirical evidence which touches every man closely. (Freud 1973, 326)

Althusser misquoted Freud and substituted Marx for Darwin when he argued:

> Not in vain did Freud sometimes compare the critical reception of his discovery with the upheavals of the Copernican Revolution. Since Copernicus, we have known that the earth is not the 'centre' of the universe. Since Marx, we have known that the human subject, the economic, political or philosophical ego is not the 'centre' of history – [...] that history has no centre [...] In turn, Freud has discovered for us that the real subject, the individual in his real essence, has not the form of the ego, centred on the 'ego', on 'consciousness' or on 'existence' – [...] that the human subject is de-centred, constituted by a structure which has no 'centre' either, except in the imaginary misrecognition of the 'ego', that is, in the ideological formations in which it 'recognizes' itself.

Principally the subject misrecognises itself in its narcissistic reflection in the fetishised commodity in which the consuming subject recentres itself (or more precisely finds itself recentred) as the addressee of (or is recentred by) every advertisement, every mass-produced message.

Walter Benjamin elaborated on Marx in what he called "the Copernican revolution in historical perception," in which:

> Formerly it was thought that a fixed point had been found in "what has been;" and one saw the present engaged in tentatively concentrating the forces of knowledge on this ground. Now this relation is to be over turned, and what has been is to become the dialectical reversal – the flash of awakened consciousness. Politics attains primacy over history. (Benjamin 1999, 388–389, 898)

Fascism reversed this revolution with the primacy of history over politics.[6]

To Freud and Marx as the decentrers of psychic and productive human subjectivity Freud added Darwin as the decentrer of embodied and biological human subjectivity. Since Darwin 'Man' no longer occupies, as William Godfrey-Smith (1979, 317; my emphasis) puts it, "a *biologically* privileged position." 'Man', however, continues to occupy a culturally privileged and pernicious position in the production of commodities from

[6] Walter Benjamin elaborated in his work on the Copernican revolution in historical perception and his critique of Fascism; see Giblett (2025) for further discussion.

raw materials supplied by the earth, the consumption of fetishised commodities manufactured from those materials, the expulsion of pollutants and wastes back into the earth-household, the construction of 'the greenhouse effect' and global boiling, and the endgame of the destruction of life in the Anthropobscene. These aspects are the elephants in the room of Byung-Chul Han's *Psychopolitics; Neoliberalism and New Technologies of Power* (2017). It does not discuss, as its cover blurb claims, "all the facets of neoliberal psychopolitics fueling our contemporary crisis of freedom." Environmental and psychoecological facets are overlooked, as are the monstrous and the uncanny (as we saw in an earlier chapter). Floods and fires in the age of global boiling brought about by neoliberal psychogeopathology are fossil-fueling the contemporary crisis of freedom for all living beings in the monstrous Anthropobscene.

In addition, *Psychopolitics'* claim that "neoliberalism discovered the productive force of the psyche" seems exaggerated (as we also saw). This discovery was hardly new as capitalist consumerism in the nineteenth century had already put it into practice in the arcades as Walter Benjamin (1999) argued following Marx's critique of the fetishism of commodities. While Freud, Marx and Darwin were decentring the subject psychologically, productively and biologically, Ford and Wanamaker were recentring the subject in consumption in a counter-Copernican revolution. Since Wanamaker, the leading founder of the American department store, the consuming subject has been reinstituted as the centre of his or her own consumptive world (see Leach 1993). Since Ford, the inventor not only of the assembly line and mass production, but also of mass consumption, the subject consuming the commodities of industrial technology (especially the car) is the centre of the history of modernity (however alienated the individual producer may be). The consumer is the centred subject of every commodity, the addressee of every shop window display, every advertisement, every fashion catwalk, every cinema, every film, every television set and show, every computer monitor. The consuming subject is a desiring subject whose desires are repressed and sublimated in and by semiosis and consumption. Desire for Deleuze and Guattari (1977, 26) "lacks a fixed subject; there is no fixed subject unless there is repression." And its counterpart sublimation I would add.

Besides the Copernican revolution in subjectivity brought about by Marx, Freud and Darwin a Copernican revolution in ecology was brought about by Ernst Haeckel, Peter Kropotkin, and Ellen Swallow (see Giblett 2011, chapter 3) that saw the embodied subject as not only a biological being with no special privileges, but also as the member of a community of living beings and non-living things with responsibilities. Since Aldo Leopold, the embodied subject is, or should be, a member of a biological community, not the owner of land as property or a consumer of commodities as fetishes. Humans should not only be part of a cultural community, but also good citizens of a natural community. Leopold's (1949, 204) "land ethic simply enlarges the boundaries of the community to include soils, waters, plants, and animals, or collectively: the land [...A] land ethic changes the role of *Homo sapiens* from conqueror of the land community to plain member and citizen of it."

The land ethic also changes the relationship to the land from mastery to mutuality and the role of the land from commodity to community. William Godfrey–Smith (1979, 314 and 317) comments that "land in Leopold's view is not a commodity which belongs to us, but a community to which we belong" where a community is construed as "a collection of individuals who engage in cooperative behaviour to their mutual benefit." Whereas the counter–Copernican revolution in capitalist consumption has triumphed, the Copernican revolution in land community has not yet had its day.

Darwin decentred the embodied subject as biological being, but his theory of "the survival of the fittest" and "the struggle for existence" was centred in the urban and industrial capitalist 'jungle' rather than in the tropical rainforest. Marx in a letter to Engels critiqued Darwin for, in the words of Keith Thomas (1984, 90),

> representing the natural world of the animal kingdom as one of free competition and for seeing among the beasts and plants his own English society, 'with its division of labour, competition, opening up of new markets, 'inventions' and the Malthusian 'struggle for existence.'

Yet 'free competition' is never a level–playing field on which all players are treated equally and the game is played fairly, just as 'free trade' is not fair trade. There are 'winners and losers,' and violence, as Walter Benjamin argued, is the only other means besides natural selection for Darwin by which winners win and losers lose. Darwin's biology for Walter Benjamin (1979, 133 and 1996, 237), "in a thoroughly dogmatic manner, regards violence as the only original means, besides natural selection, appropriate to all the vital ends of nature." Violence becomes the *sine qua non* of nature, but the violence Darwin saw in nature was cultural, or to be more precise, the violence of modern industrial capitalism was projected onto nature. Darwin for Theodore Roszak (1992, 155) "read the ethos of industrial capitalism into the jungle." The jungle was then used to support the ideology of industrial capitalism. "Nature was used as justification by those who needed legitimate acts of violence," as Sloterdijk (1987, 56) puts it, culminating in "the social Darwinist proclamation of competition as the supposed law of natural history" as he later points out (Sloterdijk 2013, 337). Nature confirmed culture, as it did in landscape aesthetics and the gentleman's park estate (see Giblett 2011, chapters 3 and 4).

The Social Darwinists took the law of the jungle and applied it to industrial capitalism and the urban jungle, but Darwin had already taken the law of industrial capitalism and the urban jungle and applied it to the tropical rainforest. Biologists such as Tim Low (2002, especially 112, 115 and 122) who use the Social Darwinian law in which "winners prosper and losers disappear" to describe the way in which some species survive in human modified environments whilst others become extinct, close the circle. Late in the day Low (2002, 280) at the end of his book charges that "one of the worst eco–crimes we commit is helping winners displace losers." The trope returns full circle to consume its own tail (tale) like the uroboros. Industrial capitalism was an urban jungle of the "survival of the fittest" that Darwin read into the rainforest. The urban and extra-urban jungle in which "winners prosper and losers disappear" reads Darwinism onto industrial capitalism.

Certainly there is competition in nature, but there is also cooperation. Nature can be "red in tooth and claw" as it was for Alfred, Lord Tennyson, though this projects the sadistic (especially oral sadistic) features of industrial capitalism and the urban 'jungle' on to nature. All those orally sadistic monsters (including crocodiles) of Hollywood films and American and British television are projections of industrial capitalist oral sadism and its monstrous machines that are used to justify and designed to deflect attention away from the latter. Nature is not only "red in tooth and claw" but also green in leaf and branch.

Cooperation is as much a part of nature as competition, if not more so. For Peter Kropotkin (1989, 6) "mutual aid is as much a law of animal life as mutual struggle, but [...] as a factor of evolution, it most probably has a far greater importance." Tim Flannery (1994, 15 and 84) has concurred that

> evolution in Australia is not driven solely by nature 'red in tooth and claw'. Here, a more gentle force – that of coadaptation – is important [...] It is cooperation rather than competition which has been selected for in many Australian environments.

For Kropotkin (1989, 30 and 60), "the war of each against all is not *the* law of nature [...] Life is struggle; and in that struggle the fittest survive," but the fittest (the survivors) are those who are the most cooperative, not the most competitive. The Hobbesian war of each against all was a projection of Hobbes's royalist reaction to the terrors of the English Civil War onto the state of nature in order to justify and legitimate restoration of the monarchy. The law of nature is each with all, for each is all.

The most mutually symbiotic survive. Symbiosis strictly considered only occurs between living beings; living things in order to live require other living beings. As William Trager (1970, 1) puts, "no organism lives alone." No 'man,' or woman, or child, or plant, or animal, is an island, as John Donne said. Yet for an organism to live, it not only requires other organisms, but also the life–supporting habitats or bioregions in which those other organisms live, and so in which it lives. Those bioregions for all species, however microscopic, are made up of intricately interweaving and interconnecting exchanges of matter which sustain life. So intertwined are these relationships that it is ultimately impossible to separate out an organism from its habitat. In a sense an organism is, or is the product and producer of, its habitat. This habitat, this bioregion is living, and bioregions live with other bioregions in a living whole breathing in air, rooted in earth, drinking water.

This idea of earth as living organism, rather than as dead matter, is by no means new. R. G. Collingwood (1945, 31 and 95) found it in the ancient Greeks. It is there too in the work of P. D. Ouspensky (1951, 178–191) in his key words of "animated nature," "a living earth" and "'dead–nature' lives" as "some gigantic [though not gargantuan or monstrous] organism." The work of V. I. Vernadsky (1945, 1–12), the inventor of the concept of the biosphere, is credited by Jeremy Rifkin (1991, 258) with being "the first modern acknowledgment of earth as a living organism." It may be more precise to see Vernadsky as the first *post*modern acknowledgment and Thoreau as the first modern acknowledgement. For Thoreau (1962, *VI*, 121), nearly a hundred years before Vernadsky, "the earth is all alive." For us, too, today, humans need to (re-learn to) live with the earth as living being.

The earth is living is in the poetry of Walt Whitman (1982, 363–368[7]) too, especially in 'The Song of the Rolling Earth', with much else besides that resonates):

> The earth does not withhold, it is generous enough,
> The truths of the earth continually wait, they are not so conceal'd either,
> They are calm, subtle, untransmissible by print,
> They are imbued through all things conveying themselves willingly…
>
> The earth does not argue,
> Is not pathetic, has no arrangements,
> Does not scream, haste, persuade, threaten, promise,
> Makes no discriminations, has no conceivable failures,
> Closes nothing, refuses nothing, shuts none out,
> Of all the powers, objects, states, it notifies, shuts none out.
>
> The earth does not exhibit itself nor refuse to exhibit itself…
> To her children the words of the eloquent dumb great mother never fail …
>
> I swear the earth shall surely be complete to him or her who shall be complete,
> The earth remains jagged and broken only to him or her who remains jagged and broken.
> I swear there is no greatness or power that does not emulate those of the earth,
> There can be no theory of any account unless it corroborate the theory of the earth,
> No politics, song, religion, behavior, or what not, is of account, unless it compare with
> the amplitude of the earth,
> Unless it face the exactness, vitality, impartiality, rectitude of the earth.
>
> I swear I begin to see love with sweeter spasms than that which responds love,
> It is that which contains itself, which never invites and never refuses.
>
> I swear I begin to see little or nothing in audible words,
> All merges toward the presentation of the unspoken meanings of the earth,
> Toward him who sings the songs of the body and of the truths of the earth,
> Toward him who makes the dictionaries of words that print cannot touch
> … the words of the earth.

[7] This poem is (strangely and inexplicably) not included in *The Portable Whitman* (1945). Perhaps its editor did not appreciate Whitman's green politics of earthly mutuality.

References

Abercrombie, M., M. Hickman, M. Johnson, and M. Thain. 1990. Symbiosis. In *The New Penguin Dictionary of Biology*, 8th ed., 542–543. London: Penguin.

Althusser, Louis. 1984. *Essays on Ideology*. Translated by Ben Brewster. London: Verso.

Benjamin, Walter. 1979. *One Way Street and Other Writings*. Translated by Edmund Jephcott and Kingsley Shorter. London: New Left Books.

Benjamin, Jessica. 1990. *The Bonds of Love: Psychoanalysis, Feminism, and the Problem of Domination*. London: Virago.

Benjamin, Walter. 1996. *Selected Writings: Volume 1, 1913–1926*. Cambridge, MA: Belknap Press of Harvard University Press.

Benjamin, Walter. 1999. *The Arcades Project*. Translated by Howard Eiland. Cambridge, MA: Belknap Press of Harvard University Press.

Berry, Thomas. 1988. *The Dream of the Earth*. San Francisco: Sierra Club Books.

Collingwood, R. G. 1945. *The Idea of Nature*. London: Oxford University Press.

Deleuze, Gilles, and Felix Guattari. 1977. *Anti-Oedipus: Capitalism and Schizophrenia*. Translated by Robert Hurley, Mark Seem and Helen Lane. New York: Viking.

Dodge, Jim. 1990. Living by Life: Some Bioregional Theory and Practice. In *Home! A Bioregional Reader*, ed. Van Andruss et al., 5–12. Philadelphia: New Society.

Flannery, Tim. 1994. *The Future Eaters: An Ecological History of the Australasian Lands and People*. Chatswood, NSW: Reed.

Freud, Sigmund. 1973. *Introductory Lectures on Psychoanalysis*. Penguin Freud Library 1. Translated by James Strachey. First published in German in 1917.

Giblett, Rod. 2008. *Sublime Communication Technologies*. Basingstoke: Palgrave Macmillan.

Giblett, Rod. 2009. *Landscapes of Culture and Nature*. Basingstoke: Palgrave Macmillan.

Giblett, Rod. 2011. *People and Places of Nature and Culture*. Bristol: Intellect Books.

Giblett, Rod. 2012. Nature is Ordinary too: Raymond Williams as the Founder of Ecocultural Studies. *Cultural Studies* 26 (6): 922–933.

Giblett, Rod. 2020. *New Lives of the Saints: Twelve Environmental Apostles*. Lanham, MD: Hamilton Books.

Giblett, Rod. 2021. *Black Swan Song: Life and Work of a Wetland Writer*. Lanham, MD: Hamilton Books.

Giblett, Rod. 2025. *Waking to Nature with Thoreau and Benjamin: Or, Pscyhopolitical Ecology*. New York: Bloomsbury.

Godfrey-Smith, William. 1979. The Value of Wilderness. *Environmental Ethics* 1:309–319.

Guattari, Felix. 1989. The Three Ecologies. *New Formations* 8:131–147.
Guattari, Felix. 2000. *The Three Ecologies*. Translated by Ian Pindar and Paul Sutton. London: Athlone.
Han, Byung-Chul. 2017. *Psychopolitics: Neoliberalism and New Technologies of Power*. Translated by Erik Butler. London and New York: Verso.
Haraway, Donna. 1995. Cyborgs and Symbionts: Living Together in the New World Order. In *The Cyborg Handbook*, ed. Chris Gray, xi–xx. London: Routledge.
Inglis, Fred. 1995. *Raymond Williams*. London: Routledge.
Klein, Milton. 1980. On Mahler's Autistic and Symbiotic Phases: An Exposition and Evaluation. *Psychoanalysis and Contemporary Thought* 4 (1): 69–105.
Kropotkin, Peter. 1989. *Mutual Aid: A Factor in Evolution*. Montreal: Black Rose Books. First published in 1902.
Leach, William. 1993. *Land of Desire: Merchants, Power and the Rise of a New American Culture*. New York: Random House.
Leopold, Aldo. 1949. *A Sand County Almanac and Sketches Here and There*. New York: Oxford University Press.
Low, Tim. 2002. *The New Nature*. Camberwell: Penguin.
Mahler, Margaret. 1952. On Child Psychosis and Schizophrenia: Autistic and Symbiotic Infantile Psychoses. *The Psychoanalytic Study of the Child* 7:286–305.
Mahler, Margaret. 1968. *On Human Symbiosis and the Vicissitudes of Individuation, Volume 1, Infantile Psychosis*. New York: International Universities Press.
Mahler, Margaret. 1972. On the First Three Sub-Phases of the Separation-Individuation Process. *International Journal of Psychoanalysis* 53:333–338.
Mahler, Margaret, and M. Furer. 1966. Development of Symbiosis, Symbiotic Psychosis, and the Nature of Separation Anxiety…. *International Journal of Psychoanalysis* 47:559–560.
Mahler, Margaret, F. Pine, and A. Bergman. 1975. *The Psychological Birth of the Human Infant: Symbiosis and Individuation*. New York: Basic Books.
Margulis, Lynn. 1971. Symbiosis and Evolution. *Scientific American* 225:49–57.
Margulis, Lynn. 1981. *Symbiosis in Cell Evolution: Life and Its Environment on the Early Earth*. San Francisco: W. H. Freeman.
Margulis, Lynn. 1998. *Symbiotic Planet: A New Look at Evolution*. New York: Basic Books.
Ouspensky, P. D. 1951. *Tertium Organum: The Third Canon of Thought: A Key to the Enigmas of the World* (3rd ed.). Translated by N. Bessaraboff and C. Bragdon. New York: Alfred A. Knopf.
Plumwood, Val. 1993. *Feminism and the Mastery of Nature*. London: Routledge.
Rifkin, Jeremy. 1991. *Biosphere Politics: A New Consciousness for a New Century*. New York: Crown.
Ross, Andrew. 1994. *The Chicago Gangster Theory of Life: Nature's Debt to Society*. London: Verso.

Roszak, Theodore. 1992. *The Voice of the Earth*. London: Bantam.
Sale, Kirkpatrick. 1985. *Dwellers in the Land: The Bioregional Vision*. San Francisco: Sierra Club Books.
Sale, Kirkpatrick. 2001. There's No Place Like Home. *The Ecologist* 31 (2): 40–43.
Sanderson, Eric. 2009. *Mannahatta: A Natural History of New York City*. New York: Abrams.
Serres, Michel. 1982. *The Parasite*. Translated by Lawrence Schehr. Baltimore, MD: Johns Hopkins University Press.
Serres, Michel. 1995. *The Natural Contract*. Translated by Elizabeth MacArthur and William Paulson. Ann Arbor, MI: University of Michigan Press.
Sloterdijk, Peter. 1987. *Critique of Cynical Reason*. Translated by Michael Eldred. Minneapolis: University of Minnesota Press. First published in German in 1983.
Sloterdijk, Peter. 2008. Strong Observation: For a Space Station Philosophy. In *Native Land: Stop Eject*, ed. Raymond Depardon and Paul Virilio, 93–99. Translated by Wieland Hoban. Paris: Fondation Cartier pour l'art contemporain.
Sloterdijk, Peter. 2011. *Spheres, Volume 1: Bubbles: Microspherology*. Translated by Wieland Hoban. Los Angeles: Semiotext(e). First published in German in 1998.
Sloterdijk, Peter. 2013. *You Must Change your Life: On Anthropotechnics*. Translated by Wieland Hoban. Cambridge, UK: Polity. First published in German in 2009.
Sloterdijk, Peter. 2016. *Stress and Freedom*. Translated by Wieland Hoban. Cambridge, UK: Polity Press. First published in German in 2011.
Sloterdijk, Peter. 2018. *What Happened in the 20th Century*. Translated by Christopher Turner. Cambridge, UK: Polity Press. First published in German in 2016.
Sloterdijk, Peter. 2020. *After God*. Translated by Ian Alexander Moore. Cambridge, UK: Polity Press. First published in German in 2017.
Sloterdijk, Peter. 2024. *Out of the World*. Translated by Corey A. Dansereau and Gill Zimmermann. Stanford, CA: Stanford University Press. First published in German in 1993.
Spretnak, Charlene. 1997. *The Resurgence of the Real: Body, Nature and Place in a Hypermodern World*. Reading, MA: Addison-Wesley.
Stern, Daniel. 1985. *The Interpersonal World of the Infant: A View from Psychoanalysis and Developmental Psychology*. New York: Basic Books.
Suzuki, David with Amanda McConnell. 1997. *The Sacred Balance: Rediscovering Our Place in Nature*. St Leonards, NSW: Allen and Unwin.
Suzuki, David. 1998. *Earth Time*. St Leonards, NSW: Allen and Unwin.
Thomas, Keith. 1984. *Man and the Natural World: Changing Attitudes in England 1500–1800*. Harmondsworth: Penguin.
Thoreau, Henry David. 1962. *The Journal of Henry D. Thoreau*, Vols. I–XIV, edited by Bradford Torrey and Franci Allen. New York: Dover.
Trager, William. 1970. *Symbiosis*. New York: Van Nostrand Reinhold.

Turner, Graeme. 1990. *British Cultural Studies: An Introduction*. London: Unwin Hyman.
Vernadsky, W. I. 1945. The Biosphere and the Noösphere. *American Scientist, 33*, 1–12.
Warming, Eugene. 1909. *Oecology of Plants: An Introduction to the Study of Plant Communities*. Oxford: Clarendon Press.
Whitman, Walt. 1945. *The Portable Walt Whitman*, ed. Mark Van Doren. New York: Penguin.
Whitman, Walt. 1982. *Complete Poetry and Collected Prose*. New York: Library of America.
Williams, Raymond. 1973. *The Country and the City*. London: Chatto and Windus.
Williams, Raymond. 1984. Between Country and City. In *Second Nature*, ed. Richard Mabey, 209–219. London: Cape.
Williams, Raymond. 1985. *Towards 2000*. Harmondsworth: Penguin.
Williams, Raymond. 1989a. *The Politics of Modernism: Against the New Conformists*, ed. Tony Pinkney. London: Verso.
Williams, Raymond. 1989b. *Resources of Hope: Culture, Democracy, Socialism*, ed. Robin Gable. London: Verso.

Index[1]

[1] Note: Page numbers followed by 'n' refer to notes.

R. Giblett, *Living with Nature in the Anthropobscene*,
https://doi.org/10.1007/978-3-032-12822-5

FSC
www.fsc.org
MIX
Papier | Fördert
gute Waldnutzung
FSC® C083411

Kolibri 360 GmbH
Ferdinand-Jühlke-Straße 7
99095 Erfurt, Deutschland
produktsicherheit@kolibri360.de